Welcome to SCIENTIFICA

YOUR NAME *Daniella O'Brien*

YOUR SCHOOL *Ilford Ursuline*

Text © Peter Ellis, Lawrie Ryan, David Sang, Jane Taylor 2005
Original illustrations © Nelson Thornes Ltd 2005

The right of Peter Ellis, Lawrie Ryan, David Sang, Jane Taylor to be identified as authors of this work has been asserted by them in accordance with the Copyright, Designs and Patents Act 1988.

Published in 2005 by:
Nelson Thornes Ltd
Delta Place
27 Bath Road
CHELTENHAM
GL53 7TH
United Kingdom

05 06 07 08 /
10 9 8 7 6 5 4 3 2 1

A catalogue record for this book is available from the British Library

ISBN 0 7487 9186 8

Illustrations by Mark Draisey, Ian West, Bede Illustration
Page make-up by Wearset Ltd

Printed in Croatia by Zrinski

Introduction

This workbook contains the really essential information you need to help you do well in Year 9. The work is split into short chunks within each of the 12 Units. You can see the names of the units in the contents list.

There are regular 'CHECKPOINT' questions to make sure you understand ideas as you meet them. There are also further homework tasks to try in which you can apply your learning. These are clearly linked to each section of the Unit you are studying. You'll find loads of different styles of exercise – so you won't get bored with Scientifica!

This workbook is linked to the exciting new Scientifica textbooks. You will find a lot more detail in these textbooks, including practical activities. However, you can use these workbooks separately.

To be sure that you are well prepared for your tests at the end of Key Stage 3 this book includes extra questions revising the science you learnt in Years 7 and 8, as well as some extra SAT-style questions.

Enjoy your journey through science with Scientifica!

CONTENTS

Inheritance and selection

9A1 Variation

If you look at any group of people you will see that they are either male or female, but within the group there will be many different heights and different styles and colours of hair. All these differences are called **variation**. There are different types of variation some of which are inherited from parents and some that depend on events that have happened in our lives.

⬤ Discontinuous variation

Discontinous variations are either one thing or another, such as sex or blood group and are usually inherited from parents. The number of people with the different varieties can be shown in a bar chart.

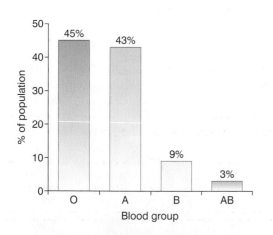

⬤ Continuous variation

Continuous variations show a gradual change from one to the other. Examples are height, weight, hair colour. They may be inherited but may also be affected by changes in the environment. For example you may have inherited the genes for tallness but you need a good diet in order to grow to your full height. You may have inherited dark hair from a parent but decide to bleach it.

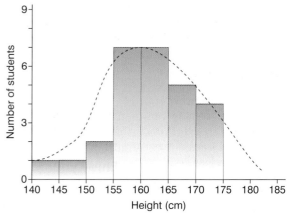

Inherited factors are controlled by **genes**. The nucleus in every cell contains the many thousands of genes that control all the different features of your body.

Ahh... you've inherited my eyes

⟩⟨ CHECKPOINT ⟩⟨

Are the variations below
a) discontinuous or continuous
b) inherited, affected by the environment or a mixture of both?

Body weight _____

Left handedness_____

Blood group _____

Length of hair _____

9A2 Passing on the genes

The features that are passed on from one generation to another are decided by the genes. Genes are carried on the **chromosomes** in the nucleus of cells. The human cell has 23 pairs of chromosomes.

The sex cells or **gametes** contain one chromosome from each pair. In animals the gametes are the egg and sperm while in plants they are the ovule and pollen.

When a sperm fertilises an egg or pollen fertilises an ovule the chromosomes pair up so that the fertilised cell has a full set of chromosomes.

The offspring inherit half of their genes from one parent and half from the other. One pair of chromosomes decides the sex of the offspring. Males have X and Y chromosomes while females have two X chromosomes in each cell. Half of the sperm made by a male will have an X chromosome and half will have a Y chromosome. The sex of the offspring depends on which sperm fertilises the egg.

Different genes

Each organism has two genes for most features. Often the two genes are different and are called **alleles**. One of the alleles is **dominant** and the other is **recessive**. The recessive feature is only **expressed**, that is appears in the organism, if both genes are recessive. The feature expressed by the dominant gene usually wins.

A genetic cross diagram shows how genes from parents result in features in the offspring. It shows the possible combinations of the two genes from each of the parents.

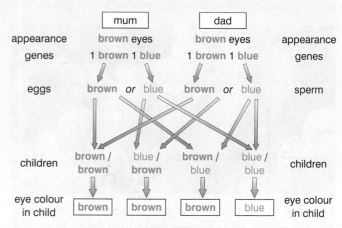

◄ CHECKPOINT ►

Draw a genetic cross diagram for the chromosomes of the possible offspring of a man (XY) and a woman (XX).

9A3 Breeding animals

For centuries humans have influenced the inheritance of domestic animals by **selective breeding**. Farmers and breeders choose the male and female animals with the features that they want to encourage. They hope that the genes for the feature will be passed on to the offspring and will be expressed.

By repeating the process over and over again they hope to produce new breeds which always have the features they selected. The selected breeds only have the alleles for the chosen features.

The Corgi has been bred for its small size

Reasons for selective breeding

- **Useful features** – features are chosen that will produce offspring with useful characteristics or skills (e.g. Jack Russell terriers, race horses)

- **Improving productivity** – features are encouraged that will give farmers more or better products to sell such as milk, meat or wool (e.g. dairy cattle, pigs, sheep)

- **Easier rearing** – features are chosen that will make the animals easier to manage. Cattle are chosen to be docile and chickens are selected that can be kept close together in sheds.

◄ CHECKPOINT ►

Why are farmyard ducks bigger than most wild varieties?

_____

How could you try to breed a greyhound to win a race?

_____

_____

_____

9A4 Breeding plants

When plants reproduce they usually produce many seeds. Each seed contains chromosomes from the male and female parent plants. The seeds may seem the same but actually have slightly different combinations of genes.

Cauliflowers, Brussels sprouts and cabbage are all the result of selective breeding from the same wild ancestors

Plant breeders make sure that the pollen from selected male plants reaches the stigma of the selected female plants. Then they repeat the process with the offspring plants that show the features they want. In this way breeders have produced many different varieties of each plant.

Breeders may select varieties that give greater productivity or are more resistant to pests or other conditions in which they are grown.

Cloning

Transferring pollen from male to female plants is a lot of hard work and some species of plant do not reproduce in this way very often. Cloning is another way of breeding plants with particular features. A bit of the plant, sometimes just a few cells, is taken and put in a nutrient bath. A new plant grows that has the same genes and features as the parent. One plant can provide many thousands of **clones**.

Genetic modification

Genes for useful features can be transferred into the cells of a plant. The modified cell can be grown into a new variety of the plant that has features that haven't appeared naturally.

◄ CHECKPOINT ►

List the ways that you could produce a new variety of apple that has a good flavour and is not attacked by pests.

_____

_____

_____

_____

9A5 Inherited conditions

Diseases such as sickle cell anaemia, cystic fibrosis, thalassaemia and haemophilia are called **genetic diseases**. They are not caused by microbes but are inherited from parents.

We have pairs of genes for thousands of different features and often the two genes in a pair are different versions or alleles of the gene. Some alleles produce features which may be harmful or not desirable. Harmful alleles are usually recessive.

Many people are **carriers** of these genes but are not aware of them because their other dominant gene is completely normal. But if a man and a woman who are both carriers for a harmful gene have children there is a chance that their child may have both genes that are harmful. The child then grows up with the genetic disease.

Coping with genetic diseases

At the moment there is no cure for genetic diseases although scientists hope that **gene therapy** will be able to correct the faulty genes in the future. Some genetic diseases can be treated to help sufferers.

Genetic testing can be used to see if someone is a carrier of a harmful gene or is likely to suffer from a genetic disease during their lives. **Genetic counselling** can help people found to be sufferers of a genetic disease. Counselling can also help parents who are carriers decide whether to have children and what may happen if a child is born with a genetic disease.

◄ CHECKPOINT ►

What is meant by

A carrier _____

_____

Genetic disease _____

_____

Genetic testing_____

_____

Genetic counselling_____

_____

Variation (9A1)

1 Molly and Benson planted some seeds along the edge of a flower bed. When the plants grew the ones at one end of the bed, under a tree, grew more slowly but had bigger leaves than the plants that were in the open. Eventually most of the plants produced red flowers but some were white. Molly and Benson measured the height of the plants when they were in flower.

Height range (cm)	Number of plants
0–5.0	0
5.1–10.0	1
10.1–15.0	4
15.1–20.0	9
20.1–25.0	5
25.1–30.0	2
30.1–35.0	0

a Sketch a graph of the variation in height.

b From the description of Molly and Benson's plants, pick out variations that are

   i) continuous: _____

   ii) discontinuous: _____

   iii) inherited: _____

   iv) affected by the environment: _____

2 Take a look at yourself and compare yourself with other members of your class and your family. Make a list of features that you have that may be inherited, affected by your lifestyle, and are continuous or discontinuous. Complete the table below:

Continuous variations	Discontinuous variations	Inherited features	Features affected by lifestyle

Passing on the genes (9A2)

1 Fill in the gaps in the following sentences.

Genes carry information about _____ that are _____ from parents. The genes are carried on _____ that are in the _____ of most cells. There is a _____ of genes for each feature. In animals the _____ and egg cells and in plants the pollen and _____ contain only one gene for each feature. When _____ takes place the genes pair up again so the offspring inherits half its _____ from each _____ but the offspring has a unique set of genes.

Words: characteristics, chromosomes, fertilisation, genes, inherited, nucleus, ovule, pair, parent, sperm.

2 In the diagram the sperm fertilising the egg carries a Y chromosome.

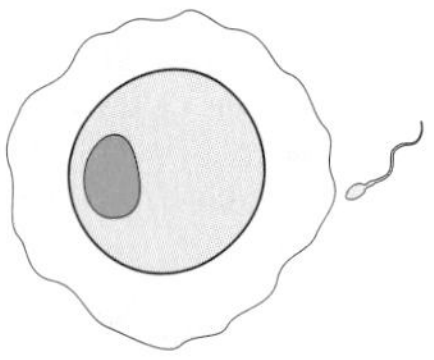

a Does the egg carry an X or a Y chromosome? Explain how you know the answer.

_____

b What will be the sex of the offspring? Explain your answer.

_____

c Y chromosomes contain some dominant genes. What does dominant mean in this case?

_____

3 Left handedness may be caused by a recessive gene.

a If a man and a woman are both left handed, will their children be left or right handed? Explain your answer.

_____

b If a left handed man had a child with a right handed woman can you say that the child would be left or right handed?

_____

Breeding animals (9A3)

1 Farmer Giles would like cows that produce milk with less fat. He chooses a cow that produces low fat milk and mates it with his bull. A female calf is born.

a When the calf grows up it does not produce low fat milk? Why is this?

_____

b What must Farmer Giles do to breed cows that produce low fat milk?

_____

2 Why have the following characteristics been bred into animals?

a Sheep that have hair that grows long before falling out.

_____

b Cattle that do not need to drink water often.

_____

c Dogs that can sense very faint smells.

_____

Breeding plants (9A4)

1. In the past new varieties of plant have been produced by pollinating selected plants or by taking cuttings from particular plants.

   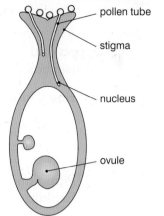

   a. Describe how a new plant is produced when pollen from one plant is dusted on to the flower of another plant.

   _____

   _____

   _____

   b. Cuttings produce plants that are *clones*. What does this mean?

   _____

   c. Fill in the table showing the advantages and disadvantages of pollination and cuttings.

Cross pollination	Taking cuttings

2. There are many different varieties of tomato. Make a list of the characteristics that may be selected for a tomato plant.

   _____

   _____

   _____

   _____

Inherited conditions (9A5)

1. A child is born with cystic fibrosis. Both parents must be carriers of the disease but neither has the disease.

   a. What is meant by *carriers*?

   _____

   b. How do we know that both parents must be carriers?

   _____

2. a. Why would a man and woman have genetic testing before choosing to have a child?

   _____

   b. How can genetic testing and counselling help parents choose whether to have children?

   _____

1 Pippa and Jamie have just had a baby daughter called Maisie. Everyone says that Maisie has Jamie's nose and Pippa's dark hair.

a) How did Maisie get her father's nose?

_____

_____

(1)

> **HINT**
> Think about sperm and egg cells.

b) Pippa and Jamie may decide to have another baby. Explain why the baby could be a boy or a girl.

_____

_____

(1)

2 Millie and Dora are identical twins but have lived apart from each other since they were born.

a) Look at the picture of Millie and Dora.

What characteristics do you think Millie and Dora inherited from their parents?

_____

_____
(2)

b) What characteristics may have been affected by Millie and Dora's environment?

_____

_____

_____

> **HINT**
> The environment means the surroundings and lifestyle.

(2)

3 Gregor grew lots of pea plants. He noticed that some plants were tall and others short. Some of the plants produced red flowers and some produced white flowers. Some of the peas were smooth and some were wrinkled.

a) Tick the box that shows the kind of variations that Gregor observed.

Continuous ☐   Discontinuous ☐   (1)

> **HINT**
> Continuous means that something can have any value across a range.

Gregor counted all the plants with red flowers and smooth peas, red flowers and wrinkled peas, white flowers with smooth peas and white flowers with wrinkled peas.

b) Tick the box naming the types of graph that Gregor should use to show his results.

Bar chart ☐   Line graph ☐   (1)

c) Underline the names of the cells that carry the genetic information when plants reproduce.

sperm   nucleus   pollen   chromosome   egg   ovule   chloroplast   (1)

d) How could Gregor produce a variety of pea plant that always produced red flowers and smooth peas?

_____

_____

_____

_____

HINT

Write out a method step by step.

(4)

e) Explain why Gregor should select the following features in his pea plants:

   i) larger peas: _____

_____

   ii) shorter stems: _____

_____

HINT

How do these features make growing the peas more profitable?

   iii) leaves that produce a substance that caterpillars dislike eating.

_____

(3)

f) Gregor finds one plant that has all the characteristics that he wants. How could he grow lots of plants with the same characteristics?

_____

_____

(1)

4 Reese and Mike have decided that they should breed mice to sell to friends as pets. They mated a brown mouse with a long tail with a brown mouse with a short tail. Some of the baby mice were brown and some white.

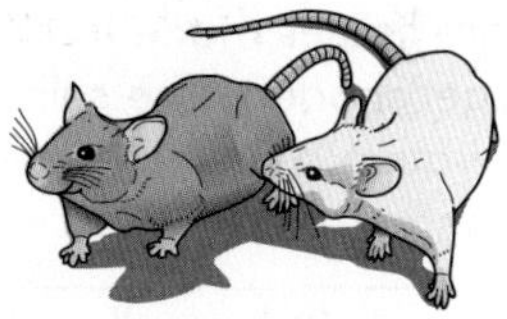

a) All the baby mice had long tails. Explain why this happened.

_____

_____ (2)

HINT

Use the terms dominant or recessive.

b) Reese and Mike decide that they want to sell white mice.

What should they do to breed white mice?

_____

_____

_____

_____

(2)

Fit and healthy

9B1 Skeleton

The bones of our skeleton do four jobs. The skeleton

- Protects organs. Your skull protects your brain and your ribs protect your lungs
- Provides support. Your spine stops all your internal organs collapsing in a heap
- Anchors muscles and gives something for muscles to pull against to make your limbs move
- Makes red blood cells. The marrow of some large bones such as in the thigh produce new red blood cells throughout your life.

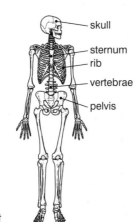

Bone material is constantly being replaced. Bone cells need a supply of calcium and vitamin D.

Exercising helps bones to grow and get stronger. When the skeleton does not have to work hard the bones become thinner and weaker.

● Joints

Bones are joined together by joints. Joints allow the bones to move. The end of the bone is covered with a layer of smooth **cartilage** which allows the bones to move easily. The two bones are connected by **ligaments**. Two types of joint are:

- **hinge joints** – such as the knee and elbow
- **ball and socket joints** – such as the hips.

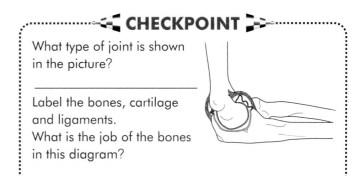

◄ CHECKPOINT ►

What type of joint is shown in the picture?

_____

Label the bones, cartilage and ligaments.
What is the job of the bones in this diagram?

_____

9B2 Muscles

Muscles are tissues made from special cells that can contract (get shorter). The cells get the energy to do this from respiration. When the muscle cells relax they become longer again.

Respiration reminder: cells get energy by reacting glucose from food with oxygen from the air.

Glucose + oxygen $\rightarrow$ carbon dioxide + water (+ energy)
$$C_6H_{12}O_6 + O_2 \rightarrow 6CO_2 + 6H_2O \ (+ \text{energy})$$

Muscles need a good blood supply to bring them the glucose and oxygen that they need.

When muscles contract they can push things. Heart muscle pushes blood into arteries; muscles in the digestive system push food through the gut.

Muscles attached to bones can pull and make the bone move.

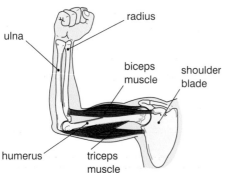

● Muscles and joints

Antagonistic pairs of muscles make a limb bend and straighten. The pairs of muscles are connected by **tendons** to the bones on each side of a joint. For example, when the biceps muscles in the upper arm contract the arm bends, when the triceps muscles contract the arm straightens.

● Joint damage

It is important to warm-up before exercising to increase the flow of blood to muscles and joints. Ligaments and tendons can be damaged and take weeks to repair themselves. The cartilage in joints can become worn if the joint is used a lot or in old age. This makes the joint swollen and painful and is called **osteoarthritis**. Joints may be replaced with artificial bones.

◄ CHECKPOINT ►

Fill in the gaps.

To make your leg bend at the knee there are a

_____ of muscles connected by _____

to the _____. When one set of muscles

_____ the leg bends. When the other muscles

contract the leg _____.

9B3 Eating well

A balanced diet provides all the nutrients we need to stay fit and healthy. We need:

- **Carbohydrates** – to provide energy
- **Proteins** – to grow new bone and muscle tissue
- **Fats** – for new cells and insulation
- **Minerals** – such as iron to make haemoglobin in red blood cells and calcium for bones
- **Vitamins** – to help cells work, such as vitamin D which bone cells need to make use of calcium
- **Fibre** – holds water as food passes through the digestive system and keeps faeces soft.

If any of the nutrients are missing from our diet we suffer from deficiency diseases and are more likely to catch infections.

Food missing from diet	Deficiency disease	Effect on health
protein	kwashiorkor	poor growth
iron	anaemia	feel tired, look pale
calcium or vitamin D	rickets	weak bones
vitamin C	scurvy	bruise easily
fibre	constipation bowel cancer	discomfort removal of part of the intestines

Malnourishment

People with foods missing from their diet or just not enough food are **malnourished**. But eating too much of some foods also causes malnourishment. Eating too much fat leads to **obesity** which can result in heart disease, diabetes and joint problems. People with **eating disorders** do not eat a balanced diet and may suffer from a number of health problems.

◄ CHECKPOINT ►

Which nutrients do you get from each of these foods

Roast chicken: _____

_____

Broccoli _____

_____

Pasta _____

_____

Sardines _____

_____

◄ CHECKPOINT ►

Why is a balanced diet important? _____

_____

_____

9B4 Unhealthy hearts

The heart beats many times a minute to pump blood to every part of the body. The muscles of the heart have their own blood vessels which supply the nutrients and oxygen to keep them working.

Circulation problems

High blood pressure means that the heart muscles have to work harder than normal. This uses up energy and can damage the heart muscles.

If we eat too much fat it can be laid down in the arteries. This makes the surface of the arteries rough. It is called an **atheroma**. A blood clot forms over the rough area and blocks the artery causing **atherosclerosis**. A blood clot in the blood vessels supplying the heart is called a **coronary thrombosis**. The muscle cells cannot get their oxygen supply and stop working. This is a heart attack and stops the blood supply to the rest of the body.

Hardening of the arteries also reduces blood flow and is called **arteriosclerosis**.

Lifestyle and the heart

- **Smoking** – carbon monoxide in the cigarette smoke replaces oxygen in red blood cells so the heart has to work faster to supply cells; nicotine increases blood pressure by making blood vessels become narrower
- **Diet** – too much fat and sugars causes atheromas to form in the arteries
- **Exercise** – regular exercise improves the blood supply to the heart muscle
- **Stress** – narrows the blood vessels increasing blood pressure.

Lack of exercise and feeling tense and stressed is not good for your heart

◄ CHECKPOINT ►

How can you keep your heart fit and healthy? _____

_____

_____

_____

9B5 Smoking

As well as affecting the heart and circulation, smoking also has effects on the breathing system.

Reminder: How do you breathe?

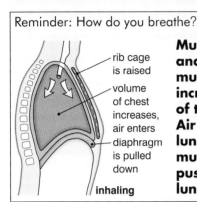

- rib cage is raised
- volume of chest increases, air enters
- diaphragm is pulled down

inhaling

Muscles across the ribs and the diaphragm muscles contract increasing the volume of the chest space. Air is pushed into the lungs. When the muscles relax air is pushed out of the lungs.

How do we know smoking harms health?

In the 1950s Richard Doll looked at the health records of thousands of smokers and non-smokers. He found that many more smokers suffered from breathing and circulation diseases than non-smokers. This showed a **correlation** between smoking and disease.

Effects of smoking

- Hot gases from cigarettes damage the ciliated cells that line the airways. The cells can no longer move mucus out of the airways so it blocks the lungs and traps microbes which cause diseases
- Hot gases and substances in the smoke kill cells in the alveoli. They are replaced with scar tissue which stops gases being exchanged. Less oxygen gets into the blood supply. This is called **emphysema**
- Substances in the tar in cigarette smoke can cause cancer in the lungs, airways or mouth
- Carbon monoxide and nicotine damage the heart
- Smoking in pregnancy can cause babies to be born prematurely or with a low birth weight
- Nicotine is addictive – it is difficult for a smoker to give up cigarettes.

◄ CHECKPOINT ►

List the organs damaged by smoking. _____

_____

_____

9B6 Alcohol and drugs

A drug is a substance that changes the way that cells in the body work. These changes can affect us physically or mentally.

Medical drugs are prescribed by doctors or pharmacists. They help us to recover from illnesses. Examples – painkillers such as aspirin or paracetamol.

Recreational drugs are the drugs that we take because we like the effect they have on us. Some drugs are present in food and drinks such as caffeine (in coffee and colas) and alcohol. Some drugs can be sold legally such as cigarettes and alcohol. Others are not available legally such as cannabis, heroin, cocaine, amphetamines and solvents.

Many drugs are **addictive**. Addicts suffer ill effects if they go without the drug. Recreational drugs may cause damage to the body and affect behaviour in ways that could cause harm.

Alcohol

Alcohol is present in beers, wines and spirits. It is absorbed into the blood through the walls of the stomach. It affects the nervous system making drinkers feel relaxed but slows down their reactions. People who have drunk alcohol are more likely to have accidents when driving or at work. More alcohol can make people aggressive and violent.

Alcohol is broken down in the liver but drinking large amounts of alcohol often damages the liver and other organs. Alcohol can pass from a pregnant woman to her baby and can cause harm to the baby.

Half a pint of beer contains about 1 unit of alcohol and a small glass of wine is $1\frac{1}{2}$ units. The liver can break down 1 unit an hour.

Men should drink no more than 21 units and women 14 units a week.

What drug is he on?

◄ CHECKPOINT ►

Make a list of the drugs that can be bought legally and illegally.

Legal	Illegal

HOMEWORK AND EXERCISES

Skeleton (9B1)

1 Write down four things that bones in the skeleton can do.

_____

_____

2 **a** Which mineral is needed in diets for healthy bones? _____

   b Which vitamin helps bone cells make use of this mineral? _____

3 Look at the diagram of a joint in the skeleton.

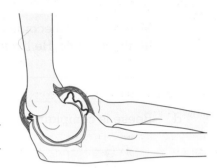

   a Label the diagram to show the position of:

   i) the ligaments

   ii) cartilage.

   b Explain what these two tissues do.

   i) ligaments: _____

   _____

   ii) cartilage: _____

   _____

Muscles (9B2)

1 Muscle cells are able to contract by using energy from respiration.

   a Write the word equation for respiration.

   _____

   b Why do muscles need a good blood supply? _____

   _____

2 Give an example of an organ in the body where muscle tissue

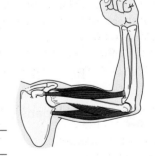

   a pushes: _____

   b pulls: _____

3 Describe what happens when you decide to straighten your arm.

   Use the terms – biceps, triceps, tendons

   _____

   _____

4 The Achilles tendon joins a muscle in the lower leg to the foot bone. Explain why people are unable to walk if they have snapped this tendon.

   _____

5 Many careers of sportspeople have been ended by cartilage damage in their knees or other joints. What does this mean?

   _____

   _____

Eating well (9B3)

1 Read the following descriptions and explain the cause of the problems. Recommend food that could help them.

    a Pip hasn't been eating very much and is complaining of feeling tired. She looks pale.

    _____

    _____

    b Marie is a young girl from a poor East European town. Her bones are bent and break often.

    _____

    _____

    c Mike wants to become a weightlifter but hasn't got the muscles to lift the weights. He likes eating sweets and crisps.

    _____

    _____

    d Reese is complaining of stomach ache and constipation. 'I haven't been for days,' she says.

    _____

    _____

    e Benson's dad has been ill and has been told he has diabetes. Benson says 'Dad has always loved a fry-up with lots of bread and butter, it's probably why he's so fat.'

    _____

    _____

Unhealthy hearts (9B4)

1 When we cut ourselves, blood clots block up the wound and prevent loss of blood.

    a What can cause blood clots inside blood vessels?

    _____

    b Why could this be dangerous?

    _____

    c Sort out this anagram to find the name of the disease.

    Roo's clear she's it. _____

2 What happens if someone has a coronary thrombosis?

_____

_____

3 How can diet, smoking and working in a stressful job affect the heart?

_____

_____

_____

_____

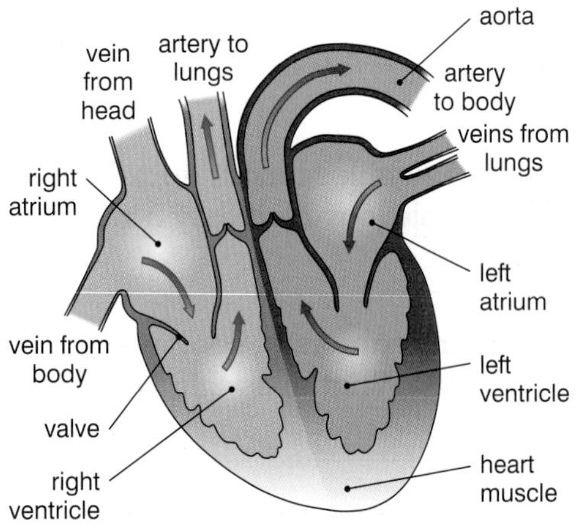

Smoking (9B5)

1 The diagram shows cells lining the trachea.

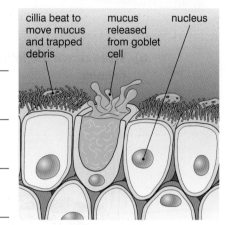

cilia beat to move mucus and trapped debris

mucus released from goblet cell

nucleus

   a What is the trachea?

_____

   b What job do the ciliated cells do?

_____

   c How does smoking affect the ciliated cells?

_____

   d What effect does this have on health?

_____

2 Match the cause to the effect.

Cause		Effect
Smoking in pregnancy		High blood pressure
Hot smoke in lungs		Low birth weight
Carbon monoxide from smoke in blood		Cancers
Nicotine from smoke in blood		Loss of surface area for gas exchange
Tar in cigarette smoke		Poor oxygen supply to tissues

Alcohol and drugs (9B6)

1 Sort these drugs into the table: penicillin, cocaine, nicotine, aspirin, caffeine, cannabis.

'Over the counter' or prescribed drugs	Legal recreational drugs	Illegal recreational drugs

2 Molly says chocolate is a drug. What evidence would you need to show that Molly is correct?

_____

3 How could you tell if a drug was addictive?

_____

4 What might be the effect of drinking the following amounts of alcohol:

   a a pint of beer

_____

   b six glasses of wine in an evening

_____

   c drinking 30 units of alcohol a week for ten years?

_____

APPLY YOUR KNOWLEDGE

1 **a)** Write down the letter in the diagram that is pointing to each of the following tissues.

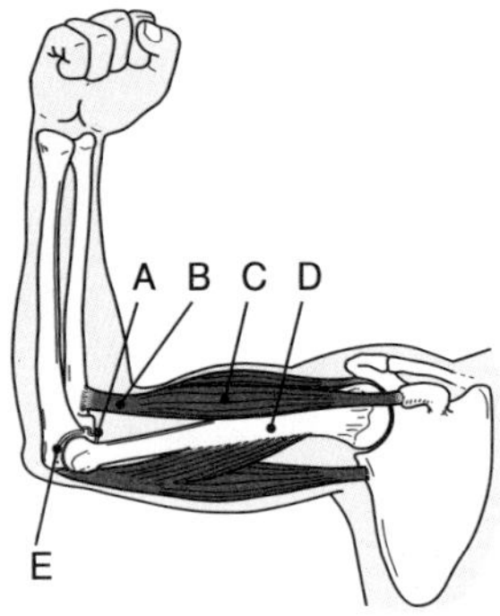

muscle ☐

tendon ☐

bone ☐

cartilage ☐

ligament ☐   (5)

HINT

Muscle cells can only contract.

b) Each joint has a pair of muscles. Explain why a pair of muscles is necessary.

_____

(1)

c) Suggest one way that a joint could be damaged in sport.

_____

(1)

2 Heart muscle is supplied with many blood vessels.

a) Why do muscle cells need a good supply of blood?

_____

(1)

HINT

Muscles use a lot of energy.

b) Carbon monoxide from cigarette smoke takes the place of oxygen in red blood cells. What effect could this have on heart muscle?

_____

(1)

c) What effect can eating too much fat and sugar have on the blood vessels in the heart muscle?

_____

(1)

HINT

Fat is not very soluble in blood.

d) Which of the following cause high blood pressure by narrowing blood vessels? Underline the correct answers:

nicotine in cigarette smoke   stress

not taking enough exercise   vitamin D deficiency   (1)

e) Meat is a good source of iron in a diet. Why is iron an essential mineral in diets?

_____

(1)

3 Use the data below to answer the following questions.

Drink	Units of alcohol
1 pint beer or lager	2
1 pint strong lager	3
1 small glass of wine	$1\frac{1}{2}$

For every unit drunk the amount of alcohol in the blood is about 40 mg/100 ml.

In 2004 the British limit for drivers is 80 mg/100 ml.

Recommended limits: men 21 units/week
women 14 units/week

a) Emily goes out for the evening with her friends. She drinks four glasses of wine.

   i) How many units of alcohol has Emily drunk? _____

   ii) How much alcohol would be in her blood? _____

   iii) Should Emily drive a car? Explain your answer. _____

   _____ (3)

HINT

Is your answer to ii) more or less than 80?

b) Jack goes to the pub every night of the week and drinks 1 pint of ordinary beer and 1 pint of strong lager.

   i) Put a ring around the number of units of alcohol he drinks in a week.

      5   15   25   35   45

   ii) Is Jack under the recommended amount for a man? _____

   iii) Describe what might happen if Jack continues with his drinking habit.

   _____

   _____ (3)

HINT

Think about the effects on his body and his behaviour.

4 'Ecstasy' became a popular but illegal recreational drug in the 1980s. The number of people who used it grew over the years. At first people thought it was safe to use but now there are concerns that it may be the cause of deaths in young people.

a) Why is Ecstasy called a drug?

_____ (1)

b) How can scientists find out if there is a correlation between the use of Ecstasy and deaths?

_____

_____ (1)

c) Why do you think the dangers of Ecstasy were not recognised in the early 1980s.

_____

_____ (1)

HINT

Think about the size of the sample.

5 Caffeine is a drug that is in many popular drinks. It can make the heart beat faster than normal.

a) Explain why caffeine is a stimulant.

_____

_____ (1)

HINT

A stimulant makes you feel you have more energy.

b) Which people should be advised not to take drinks containing a lot of caffeine?

_____ (1)

Plants and photosynthesis

9C1 Leaves

Leaves are the organs of plants that carry out **photosynthesis**.

- Leaves contain **chloroplasts** that absorb light
- They take in carbon dioxide from the air
- They draw water from the soil through the roots and stem
- They give out oxygen to the air and store waste materials.

To get as much light as possible the leaves of a plant are arranged so that they do not overlap. They are turned so that they face the source of light. If the plant is in shade the leaves will grow larger so that they can take in more light.

Leaves are thin so that gases do not have far to travel between the cells and the air.

Iodine can be used to test for starch in leaves. The iodine turns blue–black when starch is present.

The structure of leaves

- **Top layer** (the epidermis) – transparent to let light through and covered with a waxy material to stop loss of water
- **Palisade layer** – layer of cells that contain a lot of chloroplasts. It is where most photosynthesis takes place. There are veins which carry water to the cells

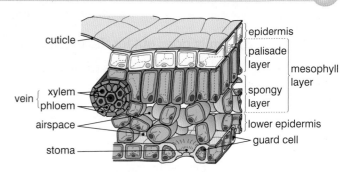

- **Mesophyll layer** – this layer has lots of spaces which acts as a store for gases
- **Bottom layer** (epidermis) – similar to the top layer but with holes, called **stomata**, surrounded by **guard cells** that control the movement of gases and water vapour.

 **CHECKPOINT**

Draw lines to link the function with the part of the plant involved.
- absorb light   palisade layer
- take in carbon dioxide from the air   stomata
- release oxygen and water vapour into the air   mesophyll layer

9C2 Photosynthesis

Photosynthesis is the process by which plants make glucose.

$$\text{Carbon dioxide} + \text{water} \xrightarrow{\text{(light)}} \text{glucose} + \text{oxygen}$$
$$6CO_2 + 6H_2O \longrightarrow C_6H_{12}O_6 + 6O_2$$

Light energy, usually from the Sun, is converted into chemical energy in the glucose. The process takes place in chloroplasts, especially in the palisade cells of leaves. Chloroplasts contain a green substance called **chlorophyll** that absorbs the light that makes the reaction work.

The oxygen that is formed in the reaction is a waste product. Most of it is released into the air but some is used by the plant's cells in respiration with the glucose.

Measuring photosynthesis

The oxygen bubbles given off by water plants such as *elodea* can be collected. The faster the bubbles are produced the greater the rate of photosynthesis.

The palisade cells

The palisade cells are the plant's specialist cells for photosynthesis. As well as containing lots of chloroplasts, their shape and position just below the upper epidermis allow them to absorb as much light as possible. Each cell is near to a supply of carbon dioxide in the air spaces of the mesophyll layer. Veins deliver water to each palisade cell.

CHECKPOINT

Link the substance used or formed in photosynthesis to where it comes from or goes.

Substance	Where it comes from or where it goes
water	from the air
carbon dioxide	released into the air
oxygen	stored in cells
glucose	absorbed from soil

9C3 Biomass

The biomass of a plant is its mass with all the water removed.

Early research into biomass

In 1648 Jean-Baptiste van Helmont found that in five years a plant grew and increased in mass. The mass of the soil in which the plant grew was unchanged. Van Helmont had watered the plant so he concluded that the extra mass of the plant came from the water.

Over a hundred years later, Joseph Priestley showed that a plant could make stale air fit to breathe again. He concluded that plants gave out the gas that we now call oxygen.

Using glucose

Glucose is used in all parts of a plant for:
- respiration, to provide energy for cell processes
- to combine with minerals to make proteins for growth and substances such as chlorophyll.
- to make other carbohydrates, for storage (starch), cell walls (cellulose) and sugars (in fruits).
- to make oils in seeds, such as olives and peanuts.

>◄ CHECKPOINT ►
>
> What was wrong about van Helmont's conclusion?
>
> _____
>
> _____

9C4 Roots

Roots anchor a plant to the ground and stop the wind and animals from knocking it over. The root system of a tree is as large as its branches.

Roots take in water and minerals from the soil. There are two types of root that do this job:
- **Taproots** are a long single root that reach deep underground to find water
- **Fibrous roots** branch into smaller and thinner roots that spread out over a large area just under the surface to collect rainwater.

Roots do not contain chloroplasts. They receive glucose from the leaves and absorb oxygen from the soil.

Root hairs

Root hair cells are special cells close to the tip of roots. They have a large surface area to take in water and minerals (e.g. nitrates) from the soil.

Water and minerals cross the root to the xylem. The xylem is a tube that runs up the middle of the root and stem to the leaves. Some of the water is used in photosynthesis but a lot evaporates from the leaves and through the stomata.

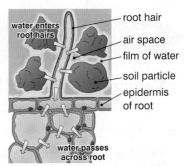

root hair
water enters root hairs
air space
film of water
soil particle
epidermis of root
water passes across root

>◄ CHECKPOINT ►
>
> Taproots spread out close to the surface   True/False
> Root hair cells do not contain chloroplasts   True/False
> Xylem carry water to the leaves   True/False
> Root hair cells collect oxygen from leaves   True/False

9C5 Why are plants important?

An **ecosystem** is made up of plants and animals that depend on each other for the substances that keep them alive. It only needs to be given light to be able to continue for a long time.

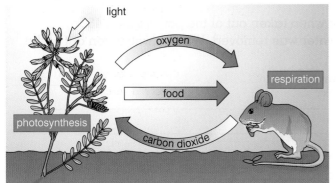

light
oxygen
respiration
food
photosynthesis
carbon dioxide

Plants are at the start of all food chains. The light energy absorbed in photosynthesis is passed from one organism to another as carbohydrates (glucose, sugars, starch and cellulose) and as proteins and oils. Plants also make the oxygen that animals need for respiration.

We depend on plants for food and fuel and use them for paper, building materials and clothes.

>◄ CHECKPOINT ►
>
> A bottle garden contains plants and small animals.
> The plants take in _____ _____
> and produce _____
> The animals take in _____
> and give out _____ _____

Leaves (9C1)

1 Fill in the gaps in the following sentences.

Photosynthesis takes place in the _____ of plants. They use energy in the form of

_____ from the Sun to make glucose. They take in _____ from the

air and draw water up from the _____ . A waste product, _____ , is released

into the _____ .

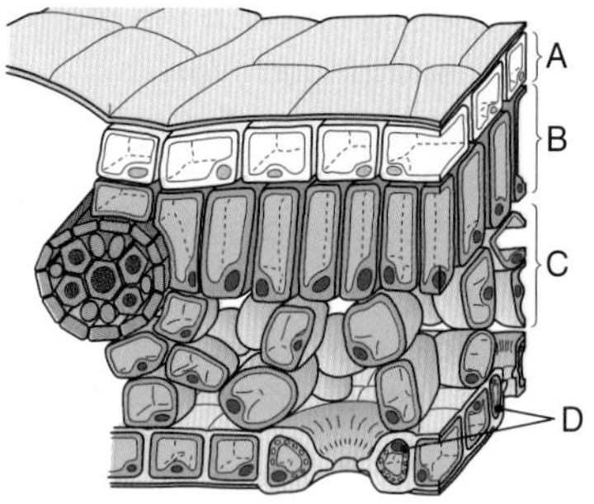

2 In which part of a leaf

  a does photosynthesis take place?

  b does water vapour and oxygen escape into the air?

  c are there air sacs that store gases?

  d does light pass through a transparent layer?

3 The carbon dioxide level in the air around a plant is lower in the daytime than at night. Why is this?

_____

Photosynthesis (9C2)

1 Fill in the gaps in the equations below

carbon dioxide + _____ → _____ + oxygen

_____ + $6H_2O$ → $C_6H_{12}O_6$ + _____

2 Name the pigment that absorbs light in leaf cells. _____

3 One plant was put in sunlight another shut in a dark box. After a few days, a leaf was taken from each plant, and placed in ethanol in a beaker of boiling water. The leaves were taken out of the ethanol and iodine solution dripped on them.

  a What makes iodine turn blue–black in colour?

  _____

  b Which leaf would have made the iodine change colour?

  _____

  c What is the conclusion of this experiment?

  _____

Let me out!
I've run out of starch

Biomass (9C3)

1 A plant in a pot of soil weighed 1.4 kg. The plant when taken out of the pot weighed 450 g. The plant was placed in an oven until all of its water was removed. When it was taken out of the oven the plant weighed 270 g.

What was the biomass of the plant? Explain your answer.

_____

2 What is glucose used for in a plant? _____

_____

Roots (9C4)

1 Root hair cells are specialised cells.

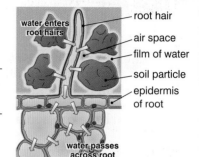

   a What do they do?

   _____

   b How are root hair cells suited to their job?

   _____

   c Why don't root hair cells contain chloroplasts?

   _____

   d Explain why root hair cells die in waterlogged soil.

   _____

   _____

2 Plants may have taproots or fibrous roots. Which type of root:

   a is made of many thin roots? _____

   b absorbs water stored deep underground? _____

   c collects rainwater? _____

3 What happens to water and minerals once it has been absorbed by root hairs?

   _____

   _____

Why are plants important? (9C5)

1 When the Earth was formed the air contained no oxygen. Now it contains 20% oxygen. What has caused this change?

   _____

   _____

2 *Biosphere 2* was an experimental habitat. It contained plants and animals including humans. No gases or materials entered or left the habitat.

   Explain how the humans were able to live in the habitat for over a year.

   _____

   _____

3 Label the arrows with the names of the substances being transferred in the diagram. Label the tree and the cow with the main processes involved in the exchange of materials.

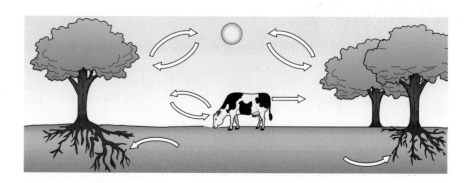

APPLY YOUR KNOWLEDGE

(SAT-style questions)

1 Mike and Reese set up the apparatus in the diagram. They placed the apparatus in a dark room with a lamp shining on it. They measured how much gas was collected in the test tube when the lamp was at different distances from the apparatus.

Here are some results that Mike and Reese collected

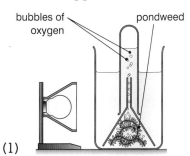

Distance of lamp from pondweed (cm)	20	40	60	80	100
Volume of gas collected (cm³)	16	4	2.5	2	1.8

a) How did Mike and Reese make sure that the results were obtained in a fair test?

_____ (1)

b) Plot a graph of Mike and Reese's results.

Plot the points and then draw a smooth line close to the points.

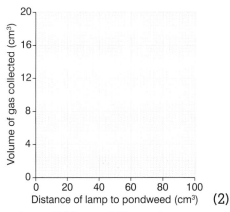

(2)

c) Underline the statements below that could have been Mike and Reese's conclusions to the experiment.

A: Pondweed gives off oxygen.

B: The further the lamp is from the pondweed the less gas is produced.

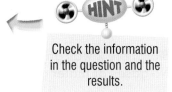

Check the information in the question and the results.

C: The speed of production of gas depends on the amount of light shining on a plant.

D: When the distance of the lamp to the pondweed is doubled the amount of gas is halved.

d) What might have happened if Mike and Reese had two lamps shining on the apparatus?

_____ (1)

e) What is the name of the process that Mike and Reese are investigating?

_____ (1)

f) Describe the process that uses light falling on a plant.

_____

_____

_____

Name the substances used and formed and where it happens.

_____ (4)

2 Palisade cells and root hair cells are special cells in plants. Complete the table comparing the two types of cells.

HINT

The special features of the cell depend on the job it does.

	Palisade cells	Root hair cells
		in tips of roots
	contain many chloroplasts	
	narrow box shape	
		absorb water and minerals from soil
	photosynthesis and respiration take place	

(5)

3 Some of the glucose made by plants is turned into cellulose.

a) What group of substances do glucose and cellulose belong to?

_____

(1)

b) Why do plants need cellulose?

_____

(1)

HINT

Think about the structure of plant cells.

c) Name one other type of substance that glucose is converted to in plants.

_____

(1)

4 Benson and Molly put two pieces of celery in a beaker of red dye solution. They left it for a few minutes. When they came back the celery with the leaves on it had red dye inside the stem.

a) What is the name of the tubes in the celery stem that the dye has risen up?

_____

(1)

b) Explain Benson and Molly's observations.

_____

_____

_____

_____

(2)

5 In the 1770s Joseph Priestley placed a mouse in a jar of air. After a few hours the mouse became unconscious. Priestley put a mint plant in the jar with another mouse. The mouse remained active. Priestley thought the mint plant was doing something to make the air breathable. Explain why animals and plants need each other.

HINT

Remember what happens in leaves.

_____

(2)

Plants for food

9D1 Plants and food

Plants use light energy to make glucose by photosynthesis.

$$\text{Carbon dioxide} + \text{water} \xrightarrow{\text{(light)}} \text{glucose} + \text{oxygen}$$

Glucose is turned into **starch** to be stored. Plants make many other substances starting with the glucose:

- **Cellulose** – for cell walls
- **Proteins** – for new cells
- **Fats and oils** – for storing energy, making waterproof layers and cells membranes
- **Vitamins** – for cell processes
- **Pigments** – for colouring flowers and leaves

◄ CHECKPOINT ►

Think of a crop. What do humans get from it? Which parts of the plants are used?

_____

_____

- **Perfumes and flavours** – to attract animals for pollination and seed dispersal
- **Protective chemicals** – to fight off unwanted animals.

We obtain many of the things we need from plants. **Crops** are plants that are grown especially for use by humans and are said to be **economically important**. Examples are:

- **Food** – grain, fruits, vegetables
- **Fibres** – such as cotton, jute, hemp, coconut are used to make ropes, cloth and paper
- **Timber** – wood from trees is used for buildings, furniture and for making paper
- **Flavours and perfumes** – tea, coffee, vanilla, mint, pine, for food and drink, cleaning and washing.

Timber... paper... chairs... floors... perfume... shelves...

Selective breeding has been used to encourage plants to produce more of the things we want and waste less of their energy on the parts that we don't use.

9D2 Fertilisers

To turn glucose into proteins, fats and the other substances, plants must take in minerals. Root hairs take in minerals dissolved in water in the soil. Plants need many different minerals but the most important are nitrates, phosphates and potassium.

Farmers spread fertiliser on their fields to provide these minerals. There are a number of fertilisers to choose from:

- Manure is decomposed animal dung. Microbes in the soil release minerals from manure slowly and it helps hold water in the soil
- Compost is similar to manure but made from rotted plant material
- 'Green manure' is plants such as clover or alfalfa. They can be grown in fields and then ploughed into the soil before planting with crops
- Artificial fertilisers are manufactured substances that dissolve in the water in the soil. They contain the minerals needed in the correct proportions.

● Fertiliser problems

If it rains after fertiliser has been spread on a field, the minerals can be washed out into rivers and lakes. This is a waste of money for the farmer but it can also harm animals in the water because it encourages growth of water plants which die and rot.

◄ CHECKPOINT ►

What happens to a plant if it is unable to get the minerals it needs from the soil?

_____

_____

9D3 Growing more crops

The **yield** of a crop is the mass of useful materials collected from a field or forest at harvest time.

To get the best yield possible from their crops farmers want to make sure that the plants grow. Their crops need to get as much light and other materials as possible for photosynthesis and other processes.

Weeds are plants that grow alongside crops but have not been planted deliberately. Weeds may grow higher than the crop and block out light or they may take water and minerals from the soil that the crop needs. The weed stops the crop from photosynthesising as fast as it could and so the crop makes less starch and other materials. The yield is reduced.

Getting rid of weeds

Farmers would like their fields to be free of weeds. The weed plants can be pulled out by hand but often farmers use **herbicides**. Herbicides are toxic chemicals that kill weeds. Weeds are often broad-leaved plants while crops such as wheat are grassy plants. The herbicides do not harm grassy plants.

Destroying weeds harms the wildlife in the area. The weeds are food for many different herbivores including molluscs such as slugs, and insects such as caterpillars. These animals are preyed on by birds and other animals. If the weeds are removed the whole food web can break down leaving no food for top predators. The populations of many types of animal will drop.

Poppies are weeds that provide food for many animals

◄ CHECKPOINT ►

Farmers spray fields with herbicides to destroy _____ and increase their _____ of the crop but wildlife is _____ because the weeds provide _____ for animals which are _____ for predators.

9D4 Pests

We grow crops for our own use but they are also attractive to other animals that eat them for food. These animals are called pests. Insects such as aphids, caterpillars and locusts are pests.

A growing crop can provide food for many pests. The number of pests will rise causing a **population boom**. Pests reduce the yield of a crop and they may also carry microbes that cause disease in the crops.

Farmers try to control the number of pests. Often farmers use chemicals called **pesticides**. If they are designed to kill insects they are known as insecticides. An alternative is **biological control** of pests. This means using a natural predator to prey on the pest that is causing the damage to crops.

Pesticide problems

Using pesticides upsets food webs. The herbivores that attack the crops are the food of other animals. If they are destroyed the predators will starve.

Sometimes a pesticide does not produce results. It may kill a natural predator of the pest more easily than the pest. The pest population increases instead of being cut.

A few pests may be resistant to pesticides and survive to produce many more pests that are resistant.

Carnivores may eat pests that have been sprayed with pesticide. The pesticide stays in their bodies. The amount of pesticide in the bodies of animals higher up the food web increases. Pesticides are **toxic** and may harm animals.

◄ CHECKPOINT ►

Growing crops can increase pest populations	True/False
Biological control uses natural predators	True/False
Herbivores accumulate more pesticides than carnivores	True/False
Pesticides kill all pests	True/False

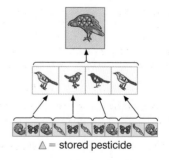

△ = stored pesticide

Gardeners and farmers try to give their plants all the conditions they need to grow successfully.

They may surround the plant with glass to trap the heat from sunlight. The extra warmth in early spring encourages seeds to germinate and young plants to grow quickly. Small glass containers are called **cloches** and large ones are **greenhouses** or **glasshouses**.

To help their plants grow gardeners will:

- cut back overgrowing plants to increase the light to their plants
- spread compost around the plants to fertilise the soil
- collect rainwater so that plants can be given water in dry spells
- remove weeds to reduce competition for minerals.

In the glasshouse

Many crops are grown in large glasshouses. Farmers can control all the conditions needed by the plants to produce a large yield. They can provide extra heat in winter and water the soil automatically. Using artificial lighting, plants can be fooled into thinking the seasons have changed. Plants can be grown all through the year providing us with flowers and fruit in all seasons.

Measuring the rate of photosynthesis

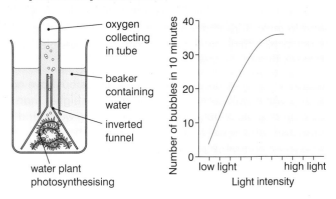

The rate of photosynthesis in various conditions can be measured using a water plant such as *elodea*. The number of oxygen bubbles produced each minute or the volume of oxygen collected over a period of time gives the rate of photosynthesis. The graph shows that as the light intensity increases the rate of photosynthesis increases.

Scientists use experiments like this to choose the best conditions for growing crops in glasshouses.

⫸ CHECKPOINT ⫷

What conditions does a plant need to grow quickly and produce a large yield?

_____

_____

_____

Plants and food (9D1)

1 Match the food to the part of the plant that it comes from. Write another food from each part alongside.

Root and tubers	celery	_____
Stem	lettuce	_____
Leaves	apple	_____
Fruit and seeds	carrot	_____

2 What are the products of photosynthesis in a plant? _____

3 Molly decided to test some parts of plants to see if they contained starch. She added iodine to samples taken from parts of the plant.

   a Which of the following would she expect to turn dark blue?

      roots   tubers (such as potato)   stem   leaves   seeds

   b What does a plant make starch for?

      _____

   c What does a plant use to make starch? _____

4 The picture shows an **economically important** food **crop** that has been produced by **selective breeding**. Explain what the highlighted terms mean.

Economically important: _____

Crop: _____

Selective breeding: _____

Fertilisers (9D2)

1 Name three minerals needed by plants.

1: _____ 2: _____ 3: _____

2 What are the advantages and disadvantages of using manure or artificial fertiliser?

Type of fertiliser	Advantages	Disadvantages
manure		
artificial		

3 The picture shows clover which farmers grow in fields in years when they do not plant a crop. Why is this a good idea?

_____

_____

4 What may happen to a plant grown on soil that has not been fertilised? Explain your answer.

_____

_____

_____

Growing more crops (9D3)

1 If **weeds** grow amongst a crop the **yield** may be reduced. Explain the highlighted terms.

Weed: _____

Yield: _____

2 Write down three ways that weeds compete with crop plants.

1: _____

2: _____

3: _____

3 PestCo have produced a new herbicide called Weedgone that kills dandelions but not grass. Pip's father sprayed his lawn with Weedgone. Later Pip noticed that there were fewer birds in the garden.

a Why would Pip's father have to wear protective clothing when spraying the lawn?

_____

b Why does Weedgone kill dandelions but not grass?

_____

c What effect would killing the dandelions have on the lawn?

_____

d Explain to Pip what has happened to the birds.

_____

Pests (9D4)

1 Pip's father has found aphids and slugs on his prize cabbages. He wants to use an insecticide to kill the pests.

a Pip explains that an insecticide will not kill the slugs. Explain why not.

_____

b Pip's father notices that the number of ladybirds on his cabbage has increased too. He wants to kill them but Pip stops him from spraying the cabbages with pesticide.

i) Why has the number of ladybirds increased?

_____

ii) Why was Pip right to stop her father spraying the cabbages?

_____

2 In many parts of the world DDT was used to spray large areas to kill malaria mosquitoes.

a In some places the number of mosquitoes fell then rose again. Why was this?

_____

b After years of spraying with DDT people noticed that there were fewer birds and small mammals. Why was this?

_____

c Top predators such as falcons do not eat mosquitoes but they were found to have high concentrations of DDT in their tissues. Why was this?

_____

1 The picture shows Pip's garden. Write on the picture four things she could do to encourage the flowers to grow.

2 Today it is easy to find glasshouse-grown strawberries in shops nearly all the year round. Explain what the growers must do to produce out of season strawberries.

_____

_____

3 What is the best temperature for photosynthesis?

Describe an experiment using pondweed to answer this question.

Sketch a diagram showing the apparatus.

_____

_____

_____

_____

_____

_____

_____

_____

_____

_____

_____

APPLY YOUR KNOWLEDGE

(SAT-style questions)

1 Mick grew some duckweed in Petri dishes containing different solutions of potassium nitrate. He counted the number of leaves in 1 cm squares and put the results in the table below.

Solution	Number of leaves in 1 cm^2					Average
	1	2	3	4	5	
distilled water	5	4	4	5	5	
1% potassium nitrate solution	8	9	8	7	8	
10% potassium nitrate solution	3	2	3	2	3	

HINT

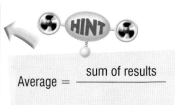

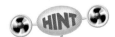

$$\text{Average} = \frac{\text{sum of results}}{}$$

a) Fill in the column of averages for each solution in the table. (3)

b) What conclusions could Mick draw from the results?

HINT

Compare the averages and the solutions used.

(1)

c) Farmers spread potassium nitrate on their fields when crops are growing. Suggest why it is important that the farmers should measure out the quantity of potassium nitrate carefully.

(1)

d) Manure is an alternative to potassium nitrate for spreading on fields. How do farmers obtain manure?

(1)

2 Winter lettuces are grown in glasshouses where powerful lights are shone on the plants day and night. The temperature is kept at about 30 °C. The plant roots are given a dilute solution of minerals.

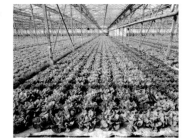

a) Which of the following substances do plants take from the air when they are in the light?

oxygen **carbon dioxide** **nitrogen** **water** (1)

b) Why do plants need light?

(1)

c) Why do the lettuces need minerals?

(1)

d) Lettuce plants usually grow in the summer. Why do glasshouse-grown lettuces grow in the winter time?

(1)

HINT

Think about how winter lettuces get the light they need.

e) Winter lettuces are usually more expensive than the summer varieties. Why is this?

(1)

3 Reese visited a farm in Norfolk where the fields were filled with lavender bushes. The lavender flowers are collected and lavender oil extracted from them for use in perfumes. Lavender oil is also an antiseptic and can be used to prevent wounds becoming infected.

a) Tick the reasons below that could explain why the plant produces lavender oil.

A: To attract bees to pollinate the plants

B: To be used to clean wounds

C: To make the flowers smell pleasant

D: To kill microbes that may try to infect the plant (1)

HINT

Plants have adaptations that help them survive.

b) The lavender bushes that Reese saw produce more oil than wild lavender bushes. Tick the reasons below that could explain this.

A: The farmed lavender bushes have been selectively bred to produce more oil.

B: The farmed lavender bushes have been given fertiliser.

C: Bees have taken all the oil from the wild lavender bushes.

D: Wild lavender bushes are not attacked by pests. (1)

c) The farmer sprays the fields with a herbicide but leaves a strip about 5 m wide around the edge of the field unsprayed.

i) What is a herbicide?

_____

HINT

Think of the effect of spraying the whole field with herbicide.

ii) Why does the farmer need to spray the crop with herbicide?

_____

iii) Why does the farmer leave the strip around the field unsprayed?

_____ (3)

d) In the spring many birds visit the fields. The birds do not eat the lavender plants so why do they visit the fields?

_____ (1)

e) Lavender is described as an economically important crop. We don't eat lavender so what does economically important mean?

_____ (1)

4 Benson used the apparatus in the diagram to measure the rate of photosynthesis of pondweed. He counted the number of bubbles produced every minute.

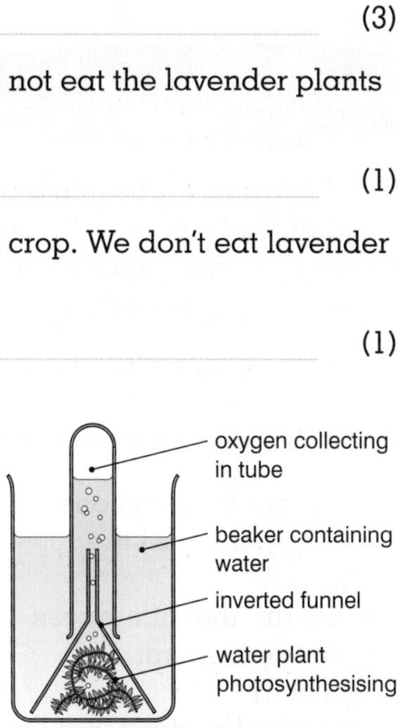

oxygen collecting in tube

beaker containing water

inverted funnel

water plant photosynthesising

a) What changes could Benson make to increase the number of bubbles produced each minute?

_____

_____ (2)

HINT

For the results to be reliable it must be a fair test.

b) What would Benson need to do to make sure the results are reliable?

_____ (1)

Cells (7A)

1 Look at the diagram of a microscope. What letter is pointing to
   a   where slides are placed
   b   the part you look through
   c   the part you hold when carrying the microscope
   d   the part used to adjust the focus of the image? 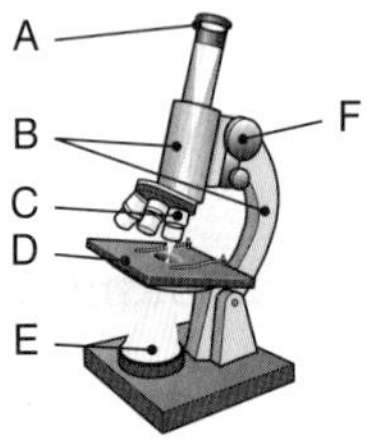
Explain how you can prepare a slide and view it through the microscope.

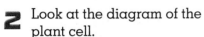

2 Look at the diagram of the plant cell. 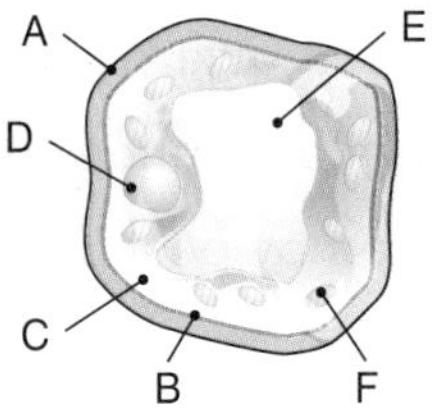
   a   Name the parts that are labelled. Explain what each part does.
   b   Which parts would you find in an animal cell?
   c   Which parts would you not find in a root hair cell? Explain your answer.

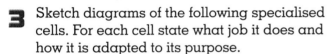

3 Sketch diagrams of the following specialised cells. For each cell state what job it does and how it is adapted to its purpose.
   a   red blood cell   d   root hair cells
   b   nerve cell   e   palisade cells.
   c   epithelial cells

4 a   What are tissues?
   b   Draw a table with the headings shown below

Organ system	Example of an organ	Specialised cells	Function

Complete the table for the circulatory system, breathing system and nervous system.

5 Describe the stages in binary fission of a cell.

6 Explain the difference between pollination and fertilisation.

7 Which of the key words are involved in the copying of cells?

KEY WORDS

adaptations   asexual   binary fission   cell
chromosome   fertilisation   microscope   nucleus
organ   pollination   slide   tissue

Reproduction (7B)

1 Put the following words into a sentence to show what they mean:
hormones   reproductive organs
adolescence   puberty

2 Put the organs in the key words into two lists, headed the female reproductive system and the male reproductive system.

3 a   What are the functions of the sperm and the egg cell?
   b   What are the differences between a sperm and egg cell?
   b   Describe the steps that lead to the fertilisation of an egg by a sperm.

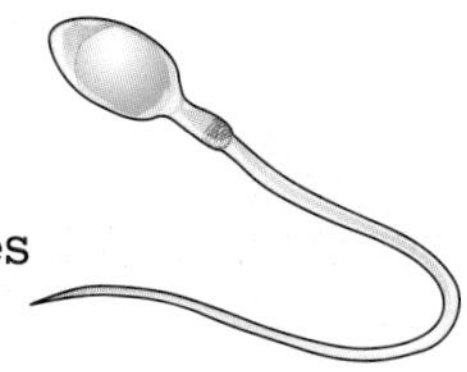

4 If an egg is fertilised a woman stops having her period.
   a   What is meant by the 'period'?
   b   What happens to the fertilised egg?

5 A foetus in the amniotic sac is connected by the umbilical cord to the placenta in the wall of the uterus. What is the purpose of
   a   the amniotic sac
   b   the placenta?

6 Why does the uterus have some of the most powerful muscles in the female body?

7 What factors could affect the health of a foetus before it is born?

8 How are the methods of reproduction of the following animals different:
   a   birds   b   mammals   c   fish.

KEY WORDS

adolescence   egg   fertilisation   hormones   implant
mammary glands   menstrual cycle   ovary   penis
placenta   pregnancy   scrotum   sperm   testes
uterus   vagina

Environment and feeding relationships (7C)

1 a Plants and animals are *adapted* to their *habitat*. What is meant by the terms habitat and adapted?

b Give some examples of adaptations of animals that live in
  i) hot, dry conditions
  ii) water.

2 Make a list of the instruments and tools that you would take on a field trip to measure and collect specimens. Say what each item would be used for.

3 a Some animals are *nocturnal*. What does this mean? How are animals adapted to being nocturnal?

b Plants and animals that live on the shoreline are adapted to the *tides*. What are the tides and how do they affect animals and plants?

4 a How do environmental conditions change during the seasons of the year?

b How do animals cope with the changes?

5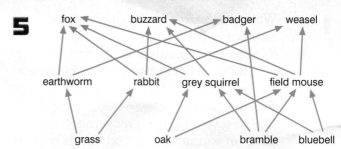

The diagram shows a food web.

a Which of the key words refer to organisms in a food web? Pick out one example from the food web.

b What adaptations may predators and prey have to help them survive?

c What factors can change the number of animals on each level?

◄ KEY WORDS ►

adapted   camouflage   community   competition
conserve   consumer   environment   food web
habitat   hibernate   migrate   nocturnal   predators
prey   producer   tides

Variation and classification (7D)

1 Which **Kingdoms** do the following organisms belong to?
  A: Contains chlorophyll and uses light to make food.
  B: Has structures that look like roots and stems but gets its food from other organisms.
  C: Can move about and gets its food from other organisms.

2 The groups of animals listed below are known as *vertebrates*. What does this mean?

  Fish   Amphibians   Reptiles
    Birds   Mammals

Describe the main features of each group.

3 Plants may be *flowering plants*, *mosses and liverworts*, *ferns*, *horsetails* or *conifers*. Which groups of plants
  a have large fronds with spores formed under their leaves?
  b produce seeds inside a fruit?

c are usually small but do not have proper roots?

d have needle-shaped leaves and produce cones?

4 In sexual reproduction two parents are needed to give genes to their offspring.
  a Why does an offspring have features of both parents?
  b How can selective breeding produce useful varieties of plants and animals?

5 Which of the key words can be used to sort organisms into groups?

◄ KEY WORDS ►

classification   flowering plant   gene   inherit
kingdom   organism   selective breeding   species
variation   vertebrate

Food and digestion (8A)

1 a Which of the key words are types of nutrient in food?

b Which type of nutrient is needed
   A: for heat insulation and making new cells
   B: for growth
   C: to give energy for movement.

c Give an example of a food that is a good source of each of the nutrients.

2 Reese ate a cheese sandwich followed by a glass of lemonade. She tested each of the foods with iodine, Benedicts solution, biuret solution and ethanol.
   a What is each substance used to test for?
   b What results would you expect Reese to get from her snack?

3 Astronauts living on the International Space Station orbiting the Earth have little fresh food to eat.
   a What may happen if the astronauts do not take vitamin tablets?
   b A vitamin tablet contains calcium and vitamin D. What are these needed for?

4 Explain the following comments about diets.
   a A child eats more protein than an adult.
   b An Inuit living in the Arctic eats more fat than someone living in the tropics.
   c An Olympic rower eats more carbohydrate than a call-centre operator.

5 Put the following organs in the correct order in the digestive system. Use key words to describe what happens in each organ.

   small intestine   mouth   stomach
   large intestine   rectum   oesophagus

6 Mike mixed some amylase solution with starch solution in a test tube and warmed it to about 40 °C. Later he added some iodine solution – it stayed yellow. He repeated the experiment at 60 °C but this time the iodine turned blue. Explain Mike's observations.

◄ KEY WORDS ►

absorption amino acids carbohydrate diet digestion enzyme fat fatty acids fibre glucose malnutrition mineral obesity protein villi vitamin

Respiration (8B)

1 Cells must be supplied with glucose and oxygen so that they can respire.
   a Why do cells need to respire?
   b What are the products of respiration?
   c How could you show that a sample of cells was respiring?

2 a How is oxygen carried in the blood?
   b The heart pumps blood through the circulatory system. Which types of blood vessel does the blood pass through in its journey around the body?
   c How do oxygen and nutrients move from the blood into cells?

3 Which of the key words are parts of the breathing system?

4 Benson decided to go for a run. Soon he was taking bigger breaths and his heart was beating faster. Then it became harder and harder to keep his legs moving and his muscles began to ache. When he stopped he carried on panting for minutes before returning to normal.

Explain what was happening to Benson.

5 Yeast use anaerobic respiration. What does this mean? Why is yeast respiration useful?

6 a Copy the diagram of the chest and label these organs:

   ribs   trachea
   bronchus   alveoli
   diaphragm

   b How is exhaled air different to normal air?

   c Describe how oxygen passes from the air into blood in an alveolus.

7 Describe how fish take in oxygen.

◄ KEY WORDS ►

aerobic alveoli breathe diaphragm diffusion energy glucose haemoglobin heart lung oxygen respire ventilate

Microbes and disease (8C)

1 Describe the characteristics of bacteria, protozoa, viruses and fungi.

2 Bacteria colonies 'grow' by increasing the number of cells.
  a How do bacteria cells multiply?
  b Describe how you could grow a colony of bacteria cells in a nutrient. Make sure you include the safety precautions.
  c Name some useful products made using bacteria or fungi.

3 Diseases are caused by microbes which can be passed from person to person.
  a Describe three ways that disease microbes can be transmitted.
  b How can infection by harmful microbes be prevented?

4 **a** How are the following used to protect the body from infection?
    A: skin
    B: tears
    C: white blood cells.

b **i)** What is antibody?
    ii) How does antibody help the fight against diseases?

5 Babies are given a *vaccine* to give them *immunity* to a disease such as measles or polio.
  a What does immunity mean?
  b Describe how the vaccine makes the baby immune to the disease.

6 Which key word describes:
  a kitchen bleach containing chlorine
  b an underarm deodorant
  c penicillin tablets?
Explain how each acts to protect us from diseases.

KEY WORDS

antibiotic antibody antigen antiseptic bacteria
colony fungi immune infection protozoa
transmission vaccine virus

Ecological relationships (8D)

1 Many *invertebrates* are *decomposers* found in soil and among dead leaves. They include *segmented worms*, *molluscs* and arthropods.
  a Explain the terms invertebrate and decomposer.
  b What are the important features of segmented worms and molluscs?
  c Where would you find decomposers in a food web?

2 **a** Why is *sampling* needed to find the number of organisms in a habitat?
  b Describe how you could sample the number of small animals in a woodland habitat.

3 **a** How would you make sure that the sample of plants in a quadrat was typical of the habitat?
  b What is a transect? Why would a transect be used?

4 Use key words to describe the effects an alien species can have on a habitat?

5 Plants compete with each other for light, water and minerals. How would a plant adapt to being shaded from sunlight?

6 Sketch a graph showing how the population of a predator, such as a buzzard, and a prey, such as a skylark, might change over a few years.

7 **a** What does the diagram show?
  b How would the diagram be changed by including parasites?

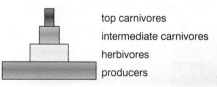

top carnivores
intermediate carnivores
herbivores
producers

KEY WORDS

alien competition decomposer distribution
parasite population pyramid of numbers quadrat
sample transect

1 The diagram shows a dandelion plant growing in long grass. When its flowers have been pollinated it produces seeds that float on the air.

a) Why is the dandelion a plant? (1)

b) Which group of plants does the dandelion belong to? (1)

c) The leaves of the dandelion contain special cells called palisade cells that contain chloroplasts.

What is the function of the palisade cells? (1)

d) Dandelions have long roots which have root hair cells on them.

What is the function of root hair cells? (1)

e) Give one difference between root hair cells and palisade cells. (1)

f) Dandelions also grow in lawns that are mown regularly. How would these dandelions be different from the one in the diagram? Explain the difference. (2)

g) A field contains hundreds of dandelions. A family of rabbits feed on the dandelions. The rabbits are eaten by a fox that lives near the field.

   i) Write a food chain for organisms in the field. (2)

   ii) Draw a pyramid of numbers for the food chain. (1)

2

Daily energy needs	(kJ/day)
Baby	4200
Child	6800
Teenage girl	9200
Adult woman	9500
Pregnant woman	10 000

a) Which nutrient provides most of our energy needs? (1)

b) Calculate the extra energy needed by an adult woman compared to a baby. (1)

c) Explain this increase. (1)

d) Why does a pregnant woman have greater energy needs? (1)

3

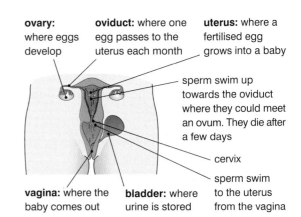

ovary: where eggs develop

oviduct: where one egg passes to the uterus each month

uterus: where a fertilised egg grows into a baby

sperm swim up towards the oviduct where they could meet an ovum. They die after a few days

cervix

sperm swim to the uterus from the vagina

vagina: where the baby comes out

bladder: where urine is stored

a) Which part of the female reproductive system releases an egg cell? (1)

b) How often is an egg cell released? (1)

c) The egg may be fertilised by a single sperm. Why are millions of sperm released into the vagina during sexual intercourse? (1)

d) What happens to the egg after it has been fertilised? (2)

4 Benson and Reese visited a cinema. Next day they started to sneeze and feel unwell. They both had colds. Three days later they began to feel better.

a) How did Benson and Reese catch colds? (1)

b) Why did it take a day before they felt unwell? (1)

c) Why did they feel better after a few days? (2)

d) Benson wanted antibiotics to cure his cold but Reese told him that there aren't antibiotics for colds. Why was Reese correct to say this? (1)

5

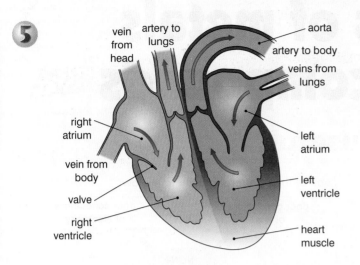

vein from head · artery to lungs · vein · aorta · artery to body · veins from lungs · right atrium · left atrium · vein from body · left ventricle · valve · right ventricle · heart muscle

a) The walls of the heart are made of muscle tissue. What does this tissue do? (1)

b) Blood vessels supply blood to the heart muscle. Why is this blood supply necessary? (2)

c) Too much fat in the diet can cause the blood vessels to become blocked. What effect may this have? (1)

Answers

1
a) They have green leaves. (1)

b) Flowering plants (1)

c) Absorb light/make sugars/photosynthesis (1)

d) To absorb water (1)

e) Root hair cells do not have chloroplasts/are not green (1)

f) Shorter stems/leaves flat on ground (1)

Less competition for light/stem does not have to grow to get light for the leaves (1)

g) i) Dandelions → rabbits → fox
correct order (1)
arrows in correct direction (1)

ii) Standard pyramid shape (1)

2
a) Carbohydrates (1)

b) 5300 kJ/day (1)

c) Larger body/more active (1)

d) Energy needed for the foetus (1)

3
a) The ovaries (1)

b) Once a month/every 28 days (1)

c) To increase the chance that a sperm will find the egg (1)

d) Divides (1)

Implants into wall of the uterus (1)

4
a) When a person with a cold sneezes droplets carry the microbe to other people nearby. (1)

b) The virus had to multiply before they felt the symptoms. (1)

c) White blood cells destroyed the viruses. (1)

But, antibody was prepared to attach to antigen on the virus to help the white blood cells destroy the virus. (1)

d) Antibiotics do not cure viral diseases. (1)

5
a) Contract to pump blood (1)

b) To provide oxygen (1)
and glucose/nutrients. (1)

c) The muscle cells die/the heart stops working/a heart attack. (1)

Reactions of metals and their compounds

9E1 Why are metals useful?

Metals

Life without metals would be difficult to imagine. We find their properties ideal for a wide variety of uses.

Look at the general properties of metals below:

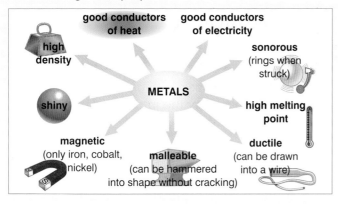

Non-metals

Less than a quarter of the elements are non-metals. Some are solids, others are gases and one, bromine, is a liquid at 20°C. Most non-metals are poor thermal and electrical conductors.

Look at their general properties below:

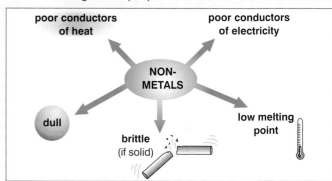

We say that 'most non-metals are poor conductors of heat and electricity'. The exception to the rule is one form of the element carbon, called **graphite**.

◄ CHECKPOINT ►

Name a property that ALL metals have.

high density

What does 'malleable' mean?

can be hammered into shape

9E2 Metals plus acids

Some metals fizz when added to dilute acids.

The reaction between a metal and any dilute acid produces hydrogen gas.

Look at the formulae of the three acids we commonly use in school:

Acid	formula
hydrochloric acid	HCl
sulphuric acid	H_2SO_4
nitric acid	HNO_3

- All acids contain hydrogen.

It is released as a gas (H_2) when a metal reacts with the acid. But what happens to the rest of the atoms in the reactants? We have the metal, of course, and the remaining atoms from the acid. These are usually formed as a solution of a **salt**.

◄ CHECKPOINT ►

Name the salt that forms when:
a) zinc reacts with dilute sulphuric acid

zinc sulphate

b) iron reacts with dilute hydrochloric acid.

iron chloride

- In general we can say:

acid + a metal → a salt + hydrogen

- A salt is a compound formed when the hydrogen in an acid is replaced (wholly or partially) by a metal.

We can show the reaction between magnesium and sulphuric acid like this:

magnesium + sulphuric → magnesium + hydrogen
acid   sulphate

Mg + H_2SO_4 → $MgSO_4$ + H_2

The salt formed is called magnesium sulphate. Its formula is $MgSO_4$.

- Sulphuric acid makes salts called **sulphates**
- Hydrochloric acid makes salts called **chlorides**
- Salts made from nitric acid are called **nitrates**.

Why can't I be the metal in the salt for a change?

9E3 Fizzing carbonates

Balancing equations

We need to balance chemical equations to make sure that we have the same number of atoms before and after a reaction. (Remember: no new atoms can be created or destroyed in a reaction – they just 'swap partners'!)

We balance an equation by inserting numbers in front of the formulae in equations, if necessary.

zinc + hydrochloric acid → zinc chloride + hydrogen
$$Zn + 2\,HCl \rightarrow ZnCl_2 + H_2$$

Notice the number 2 in front of the HCl in the symbol equation. It means we have two molecules of HCl in the equation.

This is called a **balanced symbol equation**.

We can show the particles of reactants and products using a model that represents atoms as circles. This helps us to count the atoms on each side of the equation:

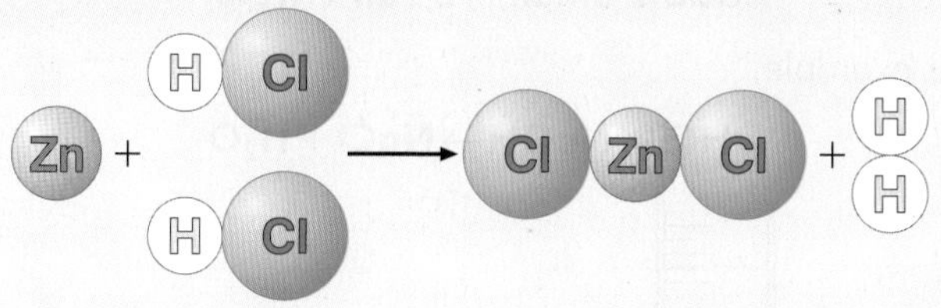

$$Zn + 2\,HCl \longrightarrow ZnCl_2 + H_2$$

Acids plus carbonates

In year 8 you came across a metal carbonate in limestone – it contains calcium carbonate, $CaCO_3$.

Copper carbonate is another metal carbonate. Its chemical formula is $CuCO_3$. When it reacts with sulphuric acid (H_2SO_4) we get:

copper + sulphuric → copper + water + carbon
carbonate   acid   sulphate   dioxide
$$CuCO_3 + H_2SO_4 \rightarrow CuSO_4 + H_2O + CO_2$$

The general equation for the reaction between a carbonate and an acid is:

acid + carbonate → a salt + water + carbon dioxide

◄ CHECKPOINT ►

a) Write the word equation for calcium carbonate reacting with hydrochloric acid.

_____

_____

b) Now write the balanced symbol equation for the same reaction.

_____

9E4 Neutralisation

Another important group of metal compounds are their oxides.

Metal oxides tend to be basic (the opposite of acidic). So they react with acids in **neutralisation** reactions. You will probably remember this from Unit 7E in *Scientifica Book 7*.

Many metal oxides do not dissolve in water. But they will dissolve in dilute acid during a neutralisation reaction.

● The metal oxide is called a **base**

● The general equation is:

acid + metal oxide (base) → a salt + water

For example:

hydrochloric acid + zinc oxide → zinc chloride + water
$$2\,HCl + ZnO \rightarrow ZnCl_2 + H_2O$$

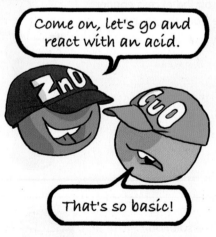

Come on, let's go and react with an acid.

That's so basic!

◄ CHECKPOINT ►

Write the word equation and balanced symbol equation for the reaction between zinc oxide and sulphuric acid.

_____

_____

_____

9E5 Preparing salts

Acids plus bases

In *Scientifica Book 9*, we saw how copper oxide (a base) neutralises an acid. We didn't need the pH sensor to tell us when the reaction was complete. We could see because the insoluble copper oxide no longer dissolved once all the acid had been neutralised.

Method for the preparation of copper sulphate

STEP 1
- Pour 20 cm³ of sulphuric acid into a small beaker. Then add a spatula of black copper oxide.
- Stir with a glass rod. Add more copper oxide, one spatula at a time until no more will dissolve.

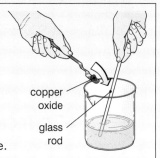

copper oxide
glass rod

STEP 2
- Warm the beaker gently on a tripod and gauze. (!DO NOT boil the mixture.)

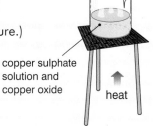

copper sulphate solution and copper oxide
heat

STEP 3
- Filter off the excess copper oxide from the solution.

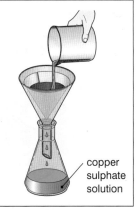

copper sulphate solution

STEP 4
- Pour the solution into an evaporating dish. Heat it on a water bath as shown.
- Stop heating when you see a few small crystals appear around the edge of the solution.

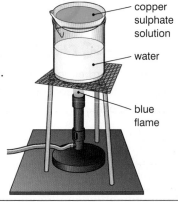

copper sulphate solution
water
blue flame

Acids plus alkalis

However, some metal oxides are soluble in water. An example is sodium oxide. So we can't tell when the sodium oxide is in excess just by looking at the reaction mixture.

In *Scientifica Book 7* you might have seen a **burette** used in a neutralisation reaction between an acid and an alkali. The burette enables us to make very small additions of one solution to another. An indicator is usually used to tell us when the reaction is just complete. The technique is called **titration**.

Alkalis are formed if we add a soluble metal oxide to water:

$$\text{sodium oxide} + \text{water} \rightarrow \text{sodium hydroxide}$$
$$Na_2O \quad + \quad H_2O \rightarrow \quad 2\,NaOH$$

Remember that:

acid + alkali → a salt + water

For example,

$HCl + NaOH \rightarrow NaCl + H_2O$

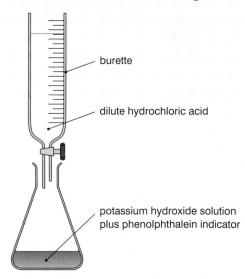

burette

dilute hydrochloric acid

potassium hydroxide solution plus phenolphthalein indicator

⟨ CHECKPOINT ⟩

Name two substances that would react to form:

a) potassium sulphate

_____

and _____

b) magnesium chloride

_____

and _____.

c) copper nitrate

_____

and _____

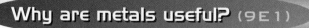

Why are metals useful? (9E1)

1 Write the word 'metal' or 'non-metal' after each of the following properties:

Low melting point _non metal_   Good electrical conductor _metal_

Good thermal conductor _metal_   Brittle _non-metal_

Malleable _metal_   High melting point _metal_

Low density _non metal_   Poor electrical conductor _non metal_

2 What do these words mean?

a sonorous _rings when struck_

b ductile _can be drawn into wire_

3 a Name a non-metal that is a good conductor of electricity _graphite_

b Name a metal that is a liquid at 20°C _mercury_

c Give one use of a particular metal and say why its properties make it good for that use.

gold is used to/for _jewlery_

because _it has a low melting point_

Metals plus acids (9E2)

1 Draw a line from the acid to its formula and then on to its salts:

sulphuric acid HNO_3 chlorides

hydrochloric acid   HCl   sulphates

nitric acid H_2SO_4 nitrates

2 a Which option below would give a positive test on the gas given off when zinc reacts with dilute sulphuric acid?

Circle the correct letter.

A: Limewater turns milky.

B: A glowing splint re-lights.

C: Damp red litmus paper turns blue.

D: Damp blue litmus paper is bleached white.

E: A lighted splint pops.

b Complete the word equation:

zinc + sulphuric acid → _zinc_ _sulphate_ + _hydrogen gas_

c Complete this balanced symbol equation:

$Zn + H_2SO_4 →$ _$ZnSo_4$_ + _H_2_

d Name one metal that does not react with dilute sulphuric acid. _____

Fizzing carbonates (9E3)

1 a Label this diagram. It shows the test for the gas given off when calcium carbonate reacts with dilute hydrochloric acid:

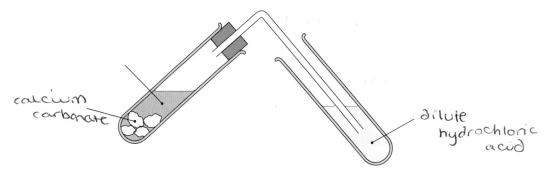

calcium carbonate

dilute hydrochloric acid

b Write down what you would **see** happen in the experiment above.

_____

_____

c Write a word equation for the reaction between calcium carbonate and dilute hydrochloric acid.

_____

d Complete the balanced symbol equation for the reaction above.

$$\underline{\hspace{3cm}} + 2\,HCl \rightarrow CaCl_2 + \underline{\hspace{3cm}} + \underline{\hspace{3cm}}$$

Neutralisation (9E4)

1. Unscramble the letters below to find the name of a base:

C R E P O P   I D O E X

The base is called _____   _____

2 Which statements are true?

Circle your answers.

A: All bases are alkalis.

B: Bases react with acids.

C: Bases react with alkalis.

D: Alkalis are soluble bases.

E: Alkalis neutralise acids.

F: Alkalis neutralise bases.

G: Acids react with alkalis to form a salt plus water.

H: Acids react with alkalis to form a salt plus hydrogen.

I: Zinc oxide is an example of a salt.

J: Zinc oxide is an example of a base.

1 You can carry out a neutralisation reaction with the help of a burette.

An indicator tells us when the reaction is just complete. This technique is called **titration**.

a Put the following steps in the correct order to carry out a titration.
The first step has been filled in for you.

A: When the indicator shows signs of changing colour, add the acid from the burette a drop at a time.

B: Collect $20\,cm^3$ of potassium hydroxide solution in a conical flask.

C: Stop when the solution turns colourless.

D: Fill a burette with dilute hydrochloric acid using a funnel.

E: Repeat the titration a few times.

F: Add a few drops of phenolphthalein solution.

G: Record how much acid is needed to neutralise the alkali.

H: Add the acid to the potassium hydroxide solution, swirling the flask as you proceed.

The correct order is:

B _____   _____   _____   _____   _____   _____   _____

b Label the diagram using information from part a.

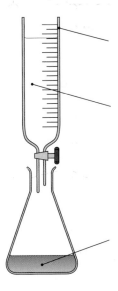

c Write the word equation for the reaction above.

_____

d Write the balanced symbol equation for the reaction above.

_____

1 Sam and Becky were preparing a sample of copper sulphate crystals. They started by reacting copper oxide with dilute sulphuric acid.

HINT

a) What do we call the chemical reaction between an acid and a base?

Acids and bases 'cancel each other out' in this type of reaction.

(1)

b) Put these steps in their method into the correct order.

HINT

Use a pencil to start sorting the steps into order. Check your order before writing your final answer in pen.

A: Pour 20 cm³ of sulphuric acid into a small beaker.

B: Pour the solution into an evaporating dish. Heat it on a water bath.

C: Warm the beaker gently on a tripod and gauze.

D: Add more copper oxide, one spatula at a time until no more will dissolve.

E: Leave the solution at room temperature for a few days.

F: Add a spatula of black copper oxide. Stir with a glass rod.

G: Stop heating when you see a few small crystals appear around the edge of the solution.

H: Filter off the excess copper oxide from the solution.

(3)

c) Write a word equation for the reaction between copper oxide and sulphuric acid.

(1)

HINT

Write down in rough the general equation for an acid plus a base. Then write your answer using the substances in the question. You will have to work out the name of the salt formed!

2 Lithium forms an oxide whose formula is Li_2O.

This lithium oxide reacts with water to form LiOH.

a) i) What is the name of LiOH?

(1)

ii) Would you expect a solution of LiOH to be acidic, neutral or alkaline?

(1)

b) i) What is the name of the salt formed when LiOH reacts with dilute nitric acid?

(1)

HINT

The metal's name comes first in the salt; then the 'back-end' comes from the acid.

ii) Write a word equation for the reaction in b) part i).

(1)

iii) LiOH also reacts with dilute hydrochloric acid.

Write a balanced symbol equation for this reaction.

(2)

3 Abbi and Kara made a salt by adding sodium carbonate powder slowly to dilute sulphuric acid.

sodium carbonate

dilute sulphuric acid

HINT
You need to know the names of common apparatus and how to use it safely.

a) What did Abbi use to add the powder to the acid?

_____ (1)

b) The method they were following told them to stir the reacting mixture. What would they use to stir the mixture?

_____ (1)

c) i) Name the two reactants in their experiment.

_____ and _____ (2)

ii) Give the name of the salt formed when the reaction was complete.

_____ (1)

iii) Write a word equation for the reaction.

_____ (1)

HINT
Think of the general equation for the reaction between a carbonate and an acid. Then write the actual reactants and products in the word equation.

d) How could Abbi and Kara see when the reaction was finished? Give two observations.

1 _____

2 _____ (2)

HINT
What happens during the reaction which will stop when all the acid has been neutralised?

e) The experiment finished with Abbi and Kara evaporating off some of the water from a solution of the salt by heating, then leaving the rest to evaporate off slowly.

Why did they let the water evaporate off slowly at the end of the experiment?

_____

_____ (1)

Patterns of reactivity

9F1 Why do metals tarnish?

Tarnished metals

When the atoms at the surface of a metal react with substances in the air, the metal loses its shiny appearance.

- We say that the metal becomes **tarnished**

- Often the metal will form a dull coating of the metal oxide as it *reacts with oxygen* in the air.

Water vapour, carbon dioxide, pollutant gases, such as sulphur dioxide, and even nitrogen can also be involved in tarnishing some metals.

This roof contains copper. The green colour comes from copper compounds which form as the copper tarnishes.

Metals react with oxygen in the air at **different rates**, if at all. Some, such as lithium and potassium, react quickly. Others react slowly, such as copper. Gold is so **unreactive** that it doesn't tarnish in air.

Gold resists tarnishing

Rusting

The rusting of iron and steel costs us millions of pounds each year.

- Iron needs both air (oxygen) and water in order to rust.

- Rust is a form of iron(III) oxide.

>-< **CHECKPOINT** >-<

Name two metals that react quickly with oxygen in the air.

_____ and _____

Name a metal that does not react with oxygen in the air.

_____

9F2 Comparing reactivity

Metals plus water

Most metals don't react vigorously with water. Just think of your pans at home! However, as with tarnishing in air, there are differences in the reactivity of different metals.

Some metals react very slowly with cold water but **react more readily with steam**.

Examples are magnesium, iron and zinc:

$$\text{zinc} + \text{steam} \rightarrow \text{zinc oxide} + \text{hydrogen}$$
$$\text{Zn} + \text{H}_2\text{O} \rightarrow \text{ZnO} + \text{H}_2$$

Reactive metals

A few metals react very quickly with cold water. As soon as they hit the water:

- they start fizzing, giving off hydrogen gas

- the solution left after the reaction is alkaline.

These alkaline solutions contain dissolved **hydroxides**. For example, lithium hydroxide, LiOH, solution.

$$\text{lithium} + \text{water} \rightarrow \text{lithium hydroxide} + \text{hydrogen}$$
$$2\,\text{Li} + 2\,\text{H}_2\text{O} \rightarrow 2\,\text{LiOH} + \text{H}_2$$

The reactive metals are from Groups 1 (and some from Group 2) in the Periodic Table.

Now that's what I call a reaction!

>-< **CHECKPOINT** >-<

Why is copper suitable for use in water pipes?

_____

_____

Write a word equation for the reaction of sodium with water.

_____

9F3 The Reactivity Series

What's the order?

We have seen how a range of metals react (or don't react) with water. We can use our observations of these reactions to place some metals into an order of reactivity.

Sorry, sir but he's been waiting a very long time for this copper to react with dilute acid!

But with some metals the reactions are very slow. This makes it difficult to put them in order of reactivity. With these metals we can look at their reactions with dilute acid to judge reactivity.

Metals of 'medium' reactivity can be added safely to dilute acid. We saw how they react with acid in Unit 9E:

$$metal + acid \rightarrow a\ salt + hydrogen$$

They fizz gently giving off hydrogen gas.

These metals include magnesium, zinc and iron. We can look at how quickly the gas is given off to put them in order.

The weird case of aluminium

Aluminium fits into the order of reactivity between magnesium and zinc. So why do we use aluminium to make cans for acidic fizzy drinks?

Aluminium is protected on its surface by a **tough layer of aluminium oxide**. Once the outer oxide layer has covered the metal, the aluminium atoms beneath are protected from water, dilute acids and oxygen.

So aluminium 'appears to be' an unreactive metal because of its oxide layer.

That's why this fairly reactive metal can be used outside, for example in patio doors. It does not corrode easily like unprotected iron.

Order of reactivity

- magnesium
- aluminium
- zinc
- iron
- copper

◄ CHECKPOINT ►

a) Which metal in the list opposite does not give off hydrogen with dilute acid?

_____

b) Which metal in the list fizzes most vigorously?

_____

9F4 Metals in competition

So far we have put the metals in order of reactivity using their reactions with water, dilute acids and oxygen. Now we can use the resulting Reactivity Series to predict what happens when we put the metals 'into competition' with each other.

For example: If we have silver nitrate solution and copper, we set up a competition between silver and copper. Both copper and silver want to form a compound. But it is silver that starts off with the nitrate as a compound in solution.

However, copper is more reactive than silver. It appears above it in the Reactivity Series. We can think of it as 'stronger' than silver. Therefore it 'takes the nitrate' for itself, going into solution to join it. The silver is 'kicked out' of solution and left as the solid element itself.

Remember that this is just a model to help us visualize what is happening. This cartoon might also help:

Hey nitrates, forget silver! I'm the more reactive one.

The word equation is:

- copper + silver nitrate → copper nitrate + silver

- We call this a **displacement reaction**.

EXAM TIP: When asked to explain this reaction, an ideal answer would be:
'Copper is **more reactive** than silver, therefore it can **displace** the silver from silver nitrate solution, forming silver metal and leaving copper nitrate in solution.'

If we add silver to copper nitrate, there will be no reaction. Silver is not reactive enough to displace copper from its solution.

◄ CHECKPOINT ►

Write the word equation for the reaction between zinc and copper sulphate solution.

_____

_____

9F5 Uses and sources of metals

Here is a table that summarises the reactions of some important metals we have met so far:

Order of reactivity	Reaction when heated in air	Reaction with water	Reaction with dilute acid
potassium	burn brightly, forming metal oxide	fizz in cold water, giving off hydrogen, leaving an alkaline solution of metal hydroxide	explode
sodium			
lithium			
calcium			fizz, giving off hydrogen and forming a salt
magnesium		no immediate reaction with cold water; react with steam, giving off hydrogen and forming the metal oxide	
aluminium			
zinc			
iron			
tin	oxide layer forms without burning	slight reaction with steam	react slowly with warm acid
lead			
copper		no reaction, even with steam	no reaction
silver	no reaction		
gold			

Look up the names of any metal you don't know in this Reactivity Series

In 9E we looked at the **physical properties** of metals. Examples include a metal's good electrical conductivity or high melting point.

In this unit we have looked at the **chemical properties** of the metals in more detail. You can link the uses, sources and methods of extraction of metals to their chemical properties.

An example is the **occurrence** of gold in nature as the metal itself.

● Only a metal very low down in the Reactivity Series could exist in nature as the element itself.

Gold can be found as the element itself because of its low reactivity. This gold nugget has a mass of over half a kilogram.

● Metals of 'medium' reactivity are extracted from ores by heating the metal oxide with carbon.

Carbon can 'take' the oxygen from any metal below aluminium in the Reactivity Series in this chemical reaction.

● We can extract more reactive metals from their ores using electrolysis.

In this process we first have to melt the metal compound, then pass electricity through it.

● So the higher up the Reactivity Series, the more difficult it gets to extract a metal from its ore.

◄ CHECKPOINT ►

a) Which metal does not burn when heated in air but does get covered by an oxide layer? This metal does not react with steam or dilute acid.

_____

b) Gold has been used for thousands of years. Why?

_____

_____

Why do metals tarnish? (9F1)

1 Look at these photos of lithium after cutting the metal with a knife:

Freshly cut lithium   **The same piece of lithium a few minutes later**

Explain what happens.

_____

_____

_____

2 a Name the two substances needed for iron to rust.

_____ and _____

b Name an iron compound that we find in rust. _____

Comparing reactivity (9F2)

1 Sort these out to make a word equation and a balanced symbol equation for the reaction of magnesium with steam:

hydrogen H_2O magnesium   MgO   water (steam) H_2 magnesium oxide   Mg

Word equation

_____

Balanced symbol equation

_____

2 a Write down three observations for the reaction of potassium with water.

i) _____

ii) _____

iii) _____

b Write a word equation for the reaction between potassium and water.

_____

c Which substance in the equation above:

i) is an alkali? _____

ii) is a metal? _____

1 Mike wanted to compare the reactivity of four metals. He had magnesium ribbon, zinc granules, iron filings and copper foil.

He decided to add them to dilute sulphuric acid. He used an electric balance to make sure he had the same mass of each metal.

Mike dropped his metals into four separate test tubes, each containing an equal volume of acid.

a Mike mixed up his observations.
Link the metal to the correct observation with a line.

Metal	Observation
zinc	no bubbles of gas given off at all
magnesium	very slowly, tiny bubbles rise through the acid
copper	after a few seconds a steady stream of small bubbles rise up
iron	fizzes quickly as soon as the metal meets the acid

b List two things that Mike did to make his test fair.

i) _____

ii) _____

c Why was Mike's experiment not really a fair test?

_____

_____

d Write a word equation for one of the reactions that Mike observed.

_____

e Put the four metals into an order of reactivity (starting with the most reactive).

1 _____ 2 _____ 3 _____ 4 _____

f Mike tried the same experiment with a piece of aluminium foil. He expected to see a steady stream of bubbles rising from the metal. However, the aluminium did not react with the acid.
Explain Mike's observations.

_____

_____

_____

Metals in competition (9F4)

1 **a** Which of these will react together?
Put a tick next to those that will react and a cross next to those that won't.

    A: iron oxide + magnesium

    B: aluminium oxide + copper

    C: copper sulphate + silver

    D: silver nitrate + iron

    E: zinc + magnesium nitrate

b Write a word equation for any reactions you expect to take place in part a.

_____

_____

_____

c What do we call this type of reaction? _____

d **i)** Explain why this reaction happens:

    iron oxide + aluminium → aluminium oxide + iron

_____

_____

    ii) Find out and explain what the reaction between iron oxide and aluminium is used for.

_____

_____

Uses and sources of metals (9F5)

1 **a** Unscramble the letters to find the names of the metals.
Then put them into the correct order of reactivity.

    L I V R E S = _____

    R O I N = _____

    U N M U L I M I A = _____

    C I N Z = _____

    Order of reactivity = 1 _____ 2 _____ 3 _____ 4 _____

            (most reactive)   (least reactive)

b Which of these metals is extracted by electrolysis? _____

c Which of these metals can be found as the element itself in nature?

_____

d Which non-metallic element is used to extract the second and third metals in the list from their ores?

_____

1. Here are some reactions of four metals:

Metal	With cold water	With dilute sulphuric acid
iron	no immediate reaction	starts fizzing, giving off a gas
silver	no reaction	no reaction
sodium	floats, then melts as it skims across the surface of the water, giving off a gas	too dangerous to attempt
nickel	no reaction	a few bubbles of gas are given off if the acid is warmed

a) List the four metals in their order of reactivity.

1 _____ (most reactive first)

2 _____

3 _____

4 _____ (1)

HINT

Use the information in the table – you don't need to remember the order of metals in the Reactivity Series – but you do need to know how to work it out.

b) i) Name another metal that reacts in a similar way to sodium with water.

_____ (1)

ii) Which gas is given off in this reaction?

_____ (1)

iii) What will be the pH of the solution that remains after the reaction?

Circle the correct answer.

1   3   7   14   (1)

iv) Why should the reaction between sodium and dilute sulphuric acid not be tried in a school laboratory?

_____ (1)

c) i) Name the gas given off when iron reacts with dilute sulphuric acid.

_____ (1)

ii) What is the other product formed in this reaction?

_____ (1)

HINT

You learnt about salts in Unit 9E. Which salt will be formed in this reaction?

d) Look at the test tubes below:

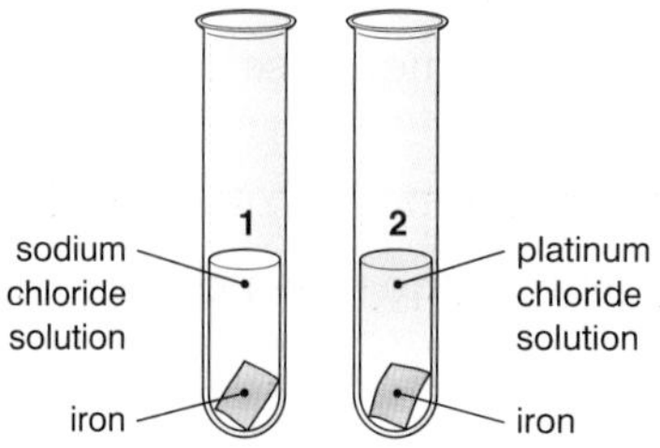

sodium chloride solution
iron
1
platinum chloride solution
iron
2

In test tube 2 the iron was slowly covered in a light grey deposit.

i) Name the light grey deposit in test tube 2.

_____ (1)

Look back to the Exam Tip on page 47 before answering this.

ii) Explain why the light grey deposit formed.

_____

_____ (1)

iii) What do we call the type of reaction observed in test tube 2?

_____ (1)

iv) Write a word equation for the reaction in test tube 2.

_____ (1)

v) Why did no reaction take place in test tube 1?

_____

_____ (1)

2 This question is about four metals – labelled A, B, C and D (these are not their chemical symbols). The metals and solutions of their sulphates were mixed on a spotting tile:

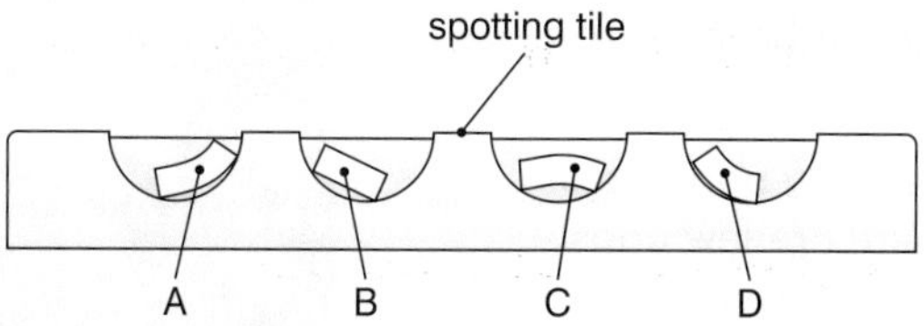

spotting tile

A   B   C   D

The table below shows the results:

✔ shows that a reaction took place   ✗ shows that no reaction took place

	A	B	C	D
A sulphate	✗	✔	✔	i)
B sulphate	✗	✗	✔	✔
C sulphate	✗	ii)	✗	iii)
D sulphate	✗	✗	✔	✗

a) Use the table to list the metals A, B, C and D in order of reactivity (most reactive first).

_____ (most reactive) _____ _____ _____ (1)

b) Enter a ✔ or a ✗ in the boxes i), ii) and iii) in the table. (2)

The most reactive metal will displace all the other metals from their solutions.

c) Copper metal reacts with silver nitrate solution.

i) Write the word equation for this reaction.

_____ (1)

ii) Platinum does not react when placed in a solution of silver nitrate.

List the three metals – platinum, copper and silver – in order of reactivity. Start with the most reactive.

_____ (1)

Environmental chemistry

9G1 Soils and weathering

Different soils

In *Scientifica Book 8* you saw how rocks can be:

- weathered (broken down)
- transported (moved from the place they were weathered)
- eroded (worn down), and then
- deposited as sediments.

These sediments form the basis of soil.

The characteristics of each type of soil is determined by:

- the size of the rock fragments it contains
- the chemical composition of the rock fragments, and
- the amount of other organic materials mixed in it. This organic material is called **humus** and comes from living organisms.

For example, a clay soil contains very tiny particles of weathered rock. This means that there are few gaps between particles for water to drain through. This type of soil gets waterlogged easily.

Compare this with a sandy soil, which feels gritty to the touch. Water drains through quickly because of the larger rock particles in the soil.

Sandy soil does have a disadvantage in that heavy rain can wash away the soluble nutrients in the soil.

We say that nutrients are **leached** from the soil.

There are different soil types and these are suited to different plants

Acidic soil

The pH of different soils can also vary. Acidic soil contains lots of organic material because it doesn't rot down in the acidic conditions. You might expect a peaty soil to be rich in nutrients because of animals and plants returning their nutrients to the soil.

However, the nutrients tend to be 'locked up' in the organic matter as it doesn't decompose easily. Therefore you need to add fertilisers. Then you get an excellent soil for growing plants as peaty soil holds moisture well.

We can neutralise acidic soil using a base, such as lime.

You only find blue hydrangeas growing naturally in acidic soil

Weathering

Weathering is the breakdown of rock by physical or chemical means. In *Scientifica Book 8* you found out about the physical effects of changes in temperature and 'freeze–thaw'.

You also looked at chemical weathering by the reaction of rocks with acidic rainwater. The process of weathering also affects building materials.

Chemical weathering will take place more quickly if you have:

- more concentrated acid. This happens in areas affected by acid rain or in the ground beneath vegetation
- high rainfall and high temperatures. These factors will also aid the breakdown of rocks by acidic solutions.

◄ CHECKPOINT ►

The sediments from different _____ produces different types of _____
We can _____ acidic soil with a basic substance such as _____ .

9G2 Acid rain

Causes of acid rain

There is a small proportion of carbon dioxide in the air. Some of it comes from *natural processes*, such as:

- volcanic activity
- respiration in living organisms, and
- the death and decomposition of living things.

This is enough to make rainwater naturally acidic. Its pH value is about 5.6.

However, pollution has made rain in many places *more acidic*. Burning **fossil fuels** adds more carbon dioxide to the air. There are also impurities of sulphur in fossil fuels.

When we burn fossil fuels, especially coal in power stations, the sulphur reacts with oxygen. It forms **sulphur dioxide gas**:

$$\text{sulphur} + \text{oxygen} \rightarrow \text{sulphur dioxide}$$
$$S + O_2 \rightarrow SO_2$$

This is a major pollution problem because sulphur dioxide is the main cause of acid rain.

The sulphur dioxide reacts with water and oxygen in the atmosphere to make **sulphuric acid**. The acidic solution falls back to the ground in rain, snow and fog.

As well as the combustion of fossil fuels, other industrial processes also give off sulphur dioxide.

Cars contribute towards the problem of acid rain. Although we can now buy fuels that have a low sulphur content, cars give off **oxides of nitrogen**. These oxides react in the atmosphere to form **nitric acid**.

Effects on building materials

We already know how dilute acids react with carbonates and some metals, dissolving them away.

- Statues lose their fine features as carbonate rock (such as marble) is slowly broken down

- Other stonework is also attacked
- Metal structures need protecting with acid-resistant paint so that corrosion does not weaken them.

Effects on plants

Alkaline soils will neutralise acid rain. However, other soils cannot neutralise much acid and essential nutrients get washed from the soil. Not only that, but acid rain attacks the waxy coating that protects leaves.

Effects on wildlife

Some water animals are very sensitive to changes in pH. They cannot survive in water whose pH value drops. Others may be more resistant, but are still affected by changes in their food chain.

- If the pH drops below 5, fish eggs will not hatch
- Once it falls below 4.5, no animal life survives.

◄ CHECKPOINT ►

Which acid forms when sulphur dioxide reacts with water and oxygen?

_____

Which acid is formed from oxides of nitrogen in the air?

_____

Where do most of the oxides of nitrogen come from?

_____

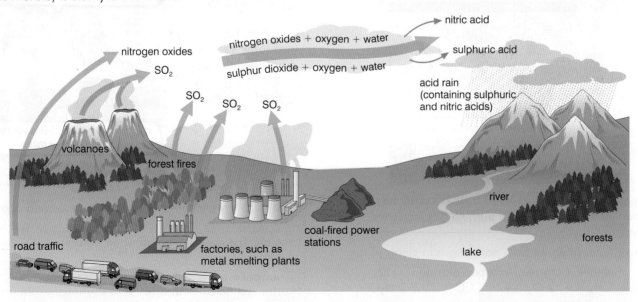

9G3 Monitoring pollution

Local Authorities are now responsible for monitoring air and water quality in their areas. You can find the data on local and national environmental websites.

Nowadays, we monitor air quality at about 1500 sites across the UK. Scientists use ever more sensitive equipment. They also use computer modelling to predict future pollution levels.

Water pollution

Rivers and lakes can become polluted by soluble substances leached from the land.

- Fertilisers get into the rivers and lakes causing **eutrophication**

- Algae thrive, cutting off light to plants on the riverbed
- The numbers of bacteria multiply as they feed on dead algae
- The bacteria use up the dissolved oxygen in the water
- This lack of oxygen kills fish and other water animals.

Water is also used as a coolant in power stations. This can cause **thermal pollution** if water is put back into the river at a higher temperature.

Oxygen gas is less soluble in warmer water, and the delicate balance of nature is disturbed.

Careful monitoring using sensitive instruments is improving the quality of rivers

◄ CHECKPOINT ►

Explain how hot water from a power station can affect fish in a river.

_____

_____

_____

9G4 Global warming

Carbon dioxide, along with water vapour, are the main 'greenhouse' gases.

The molecules of a 'greenhouse' gas absorb the heat given off by the Earth as it cools down at night. This traps the heat in the atmosphere and warms the Earth.

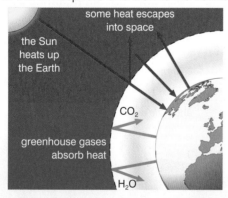

some heat escapes into space

the Sun heats up the Earth

CO_2

greenhouse gases absorb heat

H_2O

Increased carbon dioxide levels arise because the amount of fossil fuels we are burning has increased alarmingly.

For example, when we burn natural gas in plenty of air:

methane + oxygen → carbon dioxide + water

$$CH_4 + 2\,O_2 \rightarrow CO_2 + 2\,H_2O$$

Most scientists believe global warming is a serious problem and is caused by human activity. Its effects could be:

- increased sea levels and flooding of low-lying areas
- changing weather patterns all over the world. This could lead to the extinction of some species of animals. Changes in climate would also affect crop growth.

I don't fancy turkey this year!

DECEMBER 25TH

Global warming... it's not all bad, you know!

◄ CHECKPOINT ►

People are growing more and more concerned about _____ levels of _____ _____ gas in the air. The gas is given off when we burn _____ fuels.

They believe it is causing _____ warming.

Soils and weathering (9G1)

1 Rock can be broken down and eventually deposited as sediment large distances from the original rock.

Arrange these processes in order to describe 'sedimentation':

Write the number 1, 2 , 3 or 4 next to each one.

transport _____ 3 _____

deposition _____ ≠ 4 _____

weathering _____ ≠ 1 _____

erosion _____ ≠ 2 _____

2 List three ways in which soils can differ from each other.

i) _____ size _____

ii) _____ colour _____

iii) _____ shape _____

3 Explain, using a labelled diagram, why clay soils tend to get waterlogged easily but sandy soils do not.

 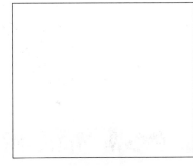

Clay soil   **Sandy soil**

Acid rain (9G2)

1 Why do fossil fuels give off sulphur dioxide gas when they burn?

Include a word equation in your answer.

_____

_____

_____

2 a Which pollutant do cars give off that contributes to acid rain?

_____

b Which acid does this make?

_____

3 Label the diagram showing how acid rain forms:

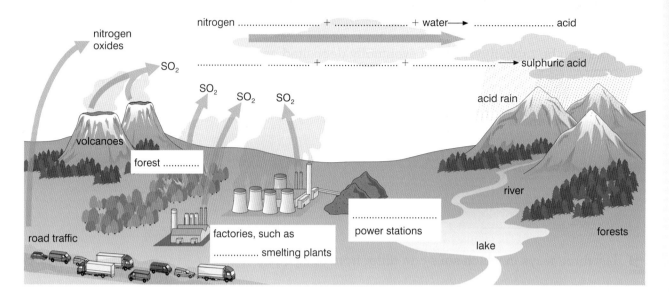

nitrogen .... + ... + water⟶ .... acid

.... + ... + ... ⟶ sulphuric acid

nitrogen oxides

SO_2

SO_2 SO_2 SO_2

acid rain

volcanoes

forest ....

river

factories, such as
.... smelting plants

.... power stations

forests

road traffic

lake

4 Find out how we can prevent the problems caused by acid rain.

_____

_____

_____

_____

_____

Monitoring pollution (9G3)

1 **a** Fertilisers can be leached from the soil into rivers.

What does the word 'leached' mean?

_____

b Once in the river the fertilisers can harm animals in the river. Put the sentences below into the correct order to describe what happens. Write the number 1, 2, 3 or 4 next to each one.

● The numbers of bacteria multiply as they feed on dead algae. _____

● The bacteria use up the dissolved oxygen in the water. _____

● This lack of oxygen kills fish and other aquatic animals. _____

● Algae flourish, cutting off light to plants on the riverbed. _____

c What do we call this pollution problem? _____

d Find out two other pollutants that also cause the problem described in part **b**.

_____ and _____

1

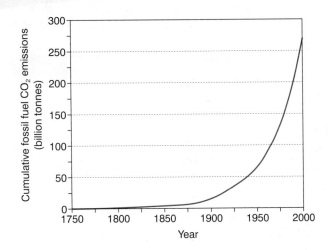

a Explain what this graph shows.

_____

_____

_____

b Why is this a problem?

_____

_____

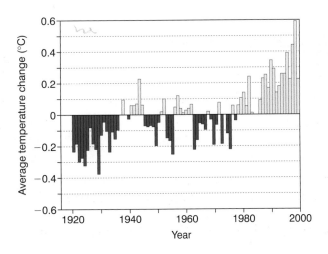

c What does this graph show?

_____

_____

d Find out why some scientists disagree with the theory that humans are causing global warming.

_____

_____

_____

APPLY YOUR KNOWLEDGE

(SAT-style questions)

1 Look at the table containing five soil samples, labelled A to E:

Soil	pH of soil
A	8.0
B	4.5
C	7.5
D	6.0
E	7.0

a) Which soil was neutral? ______ (1)

b) i) Camellia bushes grow well in acidic soils.
Which soils would be good for growing camellias?

______ and ______ (2)

ii) On the other hand lilac trees prefer alkaline conditions.
Which soils could be used to grow lilac trees?

______ and ______ (2)

HINT: Look back to Unit 7E if you still have any problems with the pH scale.

c) Farmers and gardeners sometimes add lime to acidic soil.

i) What is the chemical name for lime?

______ (1)

ii) Does lime raise or lower the pH of soil? ______ (1)

iii) What type of chemical reaction takes place between lime and the acids in soil?

______ (1)

2 Petrol contains a compound called octane, C_8H_{18}.

a) Write a word equation for the complete combustion of octane.

HINT: The carbon turns into ______ and the hydrogen turns into ______ (Remember hydrogen is never given off when a fuel burns!)

______ (1)

b) Other gases are given off from a car's exhaust. Compounds of nitrogen are formed in an engine and these contribute to acid rain.
Which compounds of nitrogen are released into the air?

______ (1)

c) Toxic carbon monoxide gas is also given off from cars. However, a catalytic converter, once it has warmed up, can change the gas into carbon dioxide.
With which environmental problem is carbon dioxide associated?

______ (1)

d) Give two ways that we can reduce the levels of carbon dioxide in the air.

______

______ (2)

In the winter of 1952, London suffered a terrible smog. (The word 'smog' came from two words – 'smoke' and 'fog'). It caused thousands of deaths.

Look at the graph below:

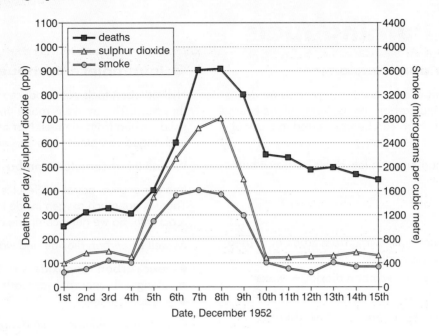

Date, December 1952

(The unit ppb of sulphur dioxide stands for 'parts per billion'.)

a) Use the graph to answer these questions:

i) How many people died on 5th December? _____ (1)

ii) On which date was there most sulphur dioxide gas in the air?

_____ (1)

iii) How much smoke was in the air on 13th December?

_____ (1)

Don't forget to include the units!

iv) By how much did the levels of smoke in the air rise between 4th December and 7th December?

_____ (1)

v) How many days did 'The Great Smog' of 1952 last? _____ (1)

How did you decide?

_____

_____

_____ (1)

Think about the kind of illnesses brought on by the smog.

b) Why do you think the death rate did not fall back to its original levels when the smog dispersed?

_____ (1)

Using chemistry

9H1 Burning fuels

Complete combustion

Lots of fuels are hydrocarbons i.e. compounds made up of only hydrogen and carbon atoms. Some of the energy released when fuels burn can be transferred into useful energy.

When a hydrocarbon burns in lots of air (so that it burns completely), we get:

● carbon dioxide, and
● water formed.

For example, propane gas, C_3H_8, from crude oil is used in some household gas heaters:

$$propane + oxygen \rightarrow carbon\ dioxide + water$$
$$C_3H_8 \quad + \quad 5\,O_2 \quad \rightarrow \quad 3\,CO_2 \quad + 4\,H_2O$$

Reactions that give out heat are called **exothermic** reactions.

So the combustion of a fuel is an example of an exothermic reaction.

Burning ethanol:

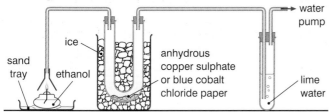

Incomplete combustion

Hydrocarbon fuels can produce some black smoke as they burn. The smoke is made up of small particles of solid carbon from the fuel. Not all the carbon in the fuel is changed completely into carbon dioxide.

We call this **incomplete combustion**.

In a car engine, petrol or diesel is ignited in a small space. There is not much oxygen inside the engine for the fuel to react with. So we get incomplete combustion of the fuel.

As well as carbon dioxide and water vapour, we also get

● toxic **carbon monoxide** gas, CO
● along with unburnt hydrocarbon fuel and
● carbon particles.

Carbon monoxide gas is so dangerous because it is odourless and effectively starves your cells of oxygen.

◄ CHECKPOINT ►

Give the products of the complete combustion of a hydrocarbon.

_____

Which gas produced by incomplete combustion of a fuel can kill you in a confined space?

_____

9H2 Energy from reactions

Chemical energy into heat energy

In Unit 9F we looked at the Reactivity Series of metals. On page 47 we saw how a more reactive metal can displace a less reactive metal. We used metals and solutions of their salts.

These **displacement** reactions, like combustion reactions, are exothermic. They release stored chemical energy as heat.

For example:

$$zinc\ sulphate + magnesium \rightarrow magnesium\ sulphate + zinc$$
$$ZnSO_4 \quad + \quad Mg \quad \rightarrow \quad MgSO_4 \quad + Zn$$

Endothermic reactions

Not all reactions are exothermic.

● Some reactions take in energy from their surroundings and the temperature falls

● These are called **endothermic** reactions. An example is photosynthesis.

Chemical energy into electrical energy

We can use the energy from displacement reactions to make **electrical cells**. The chemical energy stored can be transferred directly to electrical energy.

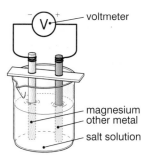

We find that:

● the greater the difference in reactivity between two metals, the larger the voltage produced.

◄ CHECKPOINT ►

The _____ energy _____ in substances can be released during d_____ reactions.
Reactions that give out heat energy are called _____ reactions.

9H3 What happens to the mass?

Particles and reactions

In Unit 9E we saw how atoms swap partners in chemical reactions. We used this to balance symbol equations. Let's look at another example below.

You know the test for hydrogen gas. You get a POP! as hydrogen and oxygen from the air react together to form water. The reaction starts when you ignite the mixture of gases. We can model the reaction as shown below:

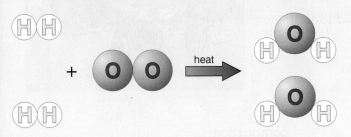

The word equation and balanced symbol equation are:

$$\text{hydrogen} + \text{oxygen} \rightarrow \text{water}$$
$$2\,H_2 + O_2 \rightarrow 2\,H_2O$$

◄ CHECKPOINT ►

If the combined mass of hydrogen and oxygen is 0.1 g (and all the molecules are converted into water), what mass of water forms in the reaction above?

_____

Conservation of mass

As no new atoms are ever created or destroyed in a chemical reaction, we can say:

the mass of reactants = the mass of products

The same thing applies in physical changes, such as melting or solidifying. The same atoms are still there before and after the change.

◄ CHECKPOINT ►

What is the chemical formula of the molecules in water and in ice? _____

If you freeze 5 g of water, what mass of ice forms?

_____

9H4 More about reactions

In *Scientifica Book 8* we looked at making compounds from elements. One of the reactions was:

$$\text{magnesium} + \text{oxygen} \rightarrow \text{magnesium oxide}$$
$$2\,Mg + O_2 \rightarrow 2\,MgO$$

We can think of the particles reacting as shown below (although the Mg atoms and the particles in MgO would really be arranged in giant structures):

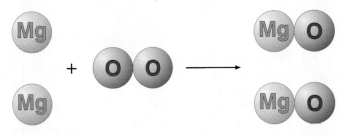

We know that the formula of magnesium oxide is MgO. The ratio of magnesium particles to oxygen particles is always 1 : 1.

We can plot a graph showing the mass of magnesium we started with and the mass of magnesium oxide it can form.

Here are some sample data:

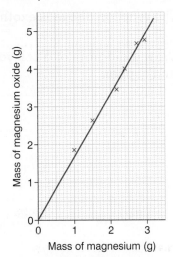

◄ CHECKPOINT ►

How much magnesium is contained in 3.0 g of magnesium oxide?

_____

HOMEWORK AND EXERCISES

Burning fuels (9H1)

1 Butane is a hydrocarbon. It is used in camping gas.

a Write a word equation for the complete combustion of butane.

_____

b Finish these sentences:

In a limited supply of oxygen, butane would form the products in part **a** as well as:

_____ and

_____

We call this _____ combustion.

c Why would burning butane in insufficient oxygen be dangerous?

_____

_____

Energy from reactions (9H2)

1 a Sort out these reactants and products into a word equation that describes a displacement reaction:

zinc sulphate   copper sulphate   zinc   copper

_____

b Delete the incorrect word:

The reaction above is **exothermic/endothermic**.

2 a Draw a diagram of a cell you could make to measure the voltage difference between zinc and copper.

b The voltage difference between zinc and copper is 1.1 volts.

Predict whether the voltage difference between magnesium and copper would be:

A: greater than 1.1 V   **B:** 1.1 V   **C:** less than 1.1 V

1 **a** Draw a particle diagram to represent this reaction:

$H_2 + Cl_2 \rightarrow 2\,HCl$

$\longrightarrow$

b What can you say about the mass of the reactants and products before and after the reaction?

_____

More about reactions (9H4)

1 Read this information about the way early scientists explained burning:

Just over 200 years ago scientists believed that substances burned if they contained a material they called **phlogiston**. They also thought that burning substances contained ash. When something burns, thought scientists in the late 18th Century, the phlogiston is given off. Then the true substance, the ash, is left behind. This was a good theory as it explained many of the observations about burning known at that time. For example:

- Hardly anything was left when charcoal (carbon) burns ('That's because it contains so much phlogiston,' said the scientists of the time.)
- We can turn the ash from a metal back into the metal by heating it with charcoal ('That's because the phlogiston from the charcoal is transferred to the ash – remember that a metal is its ash plus phlogiston!')
- Flames go out in a fixed volume of air ('That's because the air gets saturated in phlogiston and can't hold any more.').

a **i)** Imagine you were burning a piece of magnesium ribbon.
Use the phlogiston theory to explain what happens.

_____

_____

_____

ii) What is the chemical name for the 'ash' left behind after the reaction? _____

b Explain what really happens to the magnesium when it burns.
Include any equations you think will help.

_____

_____

_____

APPLY YOUR KNOWLEDGE

1 Pip and Mike heated some magnesium ribbon in a crucible, fitted with a lid.

They made sure they did not let any of the magnesium oxide formed escape.

Look at their results below:

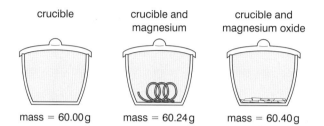

crucible	crucible and magnesium	crucible and magnesium oxide
mass = 60.00 g	mass = 60.24 g	mass = 60.40 g

a) How much magnesium did Pip and Benson start with?

_____ (1)

b) How much magnesium oxide was formed in the reaction?

_____ (1)

> **HINT**
> Use the results shown in the diagram to work these answers out.

c) Explain why the mass of the crucible and its contents increased after the reaction.

_____

_____ (1)

d) Write a word equation for the reaction.

_____ (1)

e) Balance this symbol equation for the reaction:

$$Mg + O_2 \rightarrow MgO$$
(1)

f) What is this type of chemical reaction called?

_____ (1)

> **HINT**
> Count the atoms on either side of the arrow. You can only use numbers in front of Mg, O_2 or MgO to balance the equation.

2 Molly and Pete heated some copper carbonate powder in a crucible.

They made sure no powder was lost during their experiment. They used an electric balance to find the mass before and after the experiment.

Here are their results:

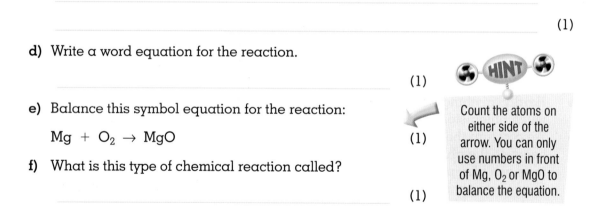

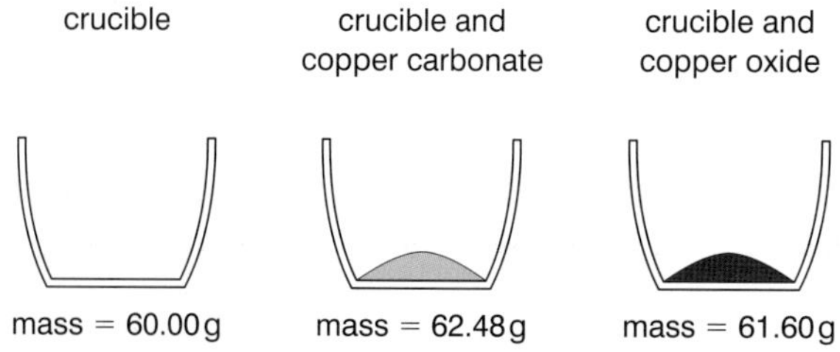

crucible	crucible and copper carbonate	crucible and copper oxide
mass = 60.00 g	mass = 62.48 g	mass = 61.60 g

a) What mass of copper carbonate did Molly and Pete start with?

_____ (1)

b) This is the word equation for the reaction:

copper carbonate → copper oxide + carbon dioxide

HINT

What will happen to the carbon dioxide produced?

   i) Why did the crucible and its contents lose mass after the reaction?

_____

_____ (1)

   ii) How much copper oxide was formed in the reaction? _____ (1)

c) What do we call this type of reaction? Choose from this list:

Underline your answer.

combustion
displacement
reduction
thermal decomposition
precipitation (1)

3 Reese and Mike did five experiments making magnesium oxide from magnesium in a crucible. They measured the mass of magnesium before the reaction and the mass of magnesium oxide formed. Here are their results on a graph:

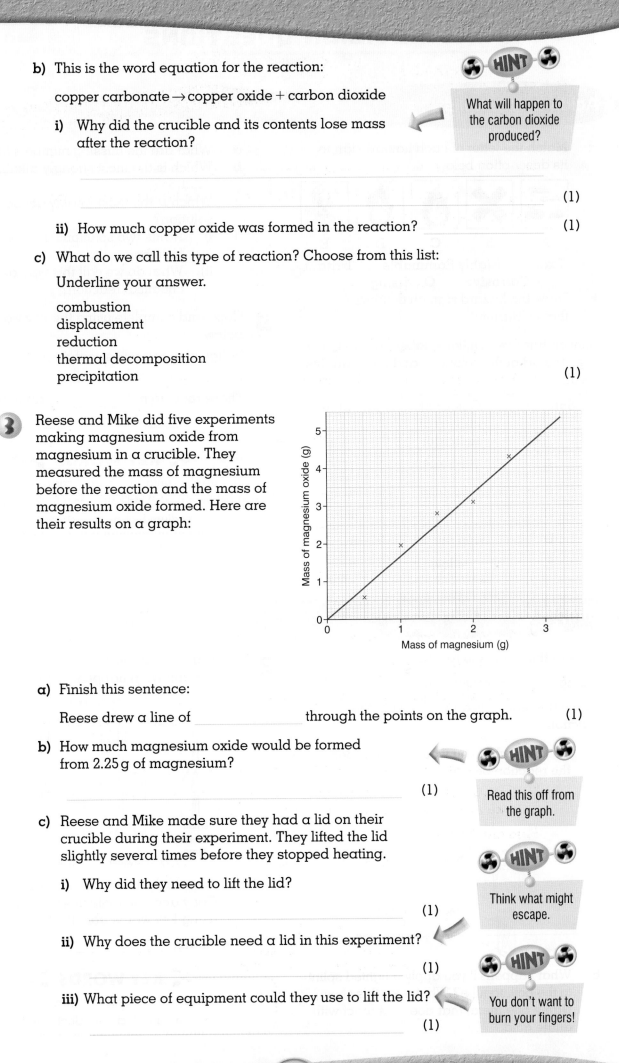

a) Finish this sentence:

Reese drew a line of _____ through the points on the graph. (1)

b) How much magnesium oxide would be formed from 2.25 g of magnesium?

_____ (1)

HINT

Read this off from the graph.

c) Reese and Mike made sure they had a lid on their crucible during their experiment. They lifted the lid slightly several times before they stopped heating.

   i) Why did they need to lift the lid?

_____ (1)

HINT

Think what might escape.

   ii) Why does the crucible need a lid in this experiment?

_____ (1)

   iii) What piece of equipment could they use to lift the lid?

_____ (1)

HINT

You don't want to burn your fingers!

Acids and alkalis (7E)

1 a Match the letter of each hazard sign to its description below.

   A   B   C   D   E

Toxic   Highly flammable   Irritant
Corrosive   Oxidising

b Draw the hazard sign on a substance that is 'Harmful'.

2 Benson has five solutions, labelled A to E. He tests the pH of the solutions and these are his results:

Solution	Colour of universal indicator	pH value
A	dark red	i
B	turquoise	ii
C	orange	iii
D	deep purple	iv
E	yellow	v

a What are the missing numbers i to v?
b Which is the most strongly alkaline solution?
c Which is the most weakly acidic solution?
d i) Name two solutions that would react together.
   ii) What do we call this type of chemical reaction?

3 Copy and complete using the key words below:

Indigestion remedies contain a
b_____ called an_____ .
These react to n_____ an excess of
a_____ in your s_____ .

<< KEY WORDS >>

acid   alkali   antacid   base   corrosive   harmful
hydrochloric acid   indicator   neutralise   pH scale
sodium hydroxide   stomach

Simple chemical reactions (7F)

1 Look at this chemical reaction:

magnesium + oxygen → magnesium oxide

Name the reactants and product in the reaction.

2 a Describe what you would see happen in the reaction below:

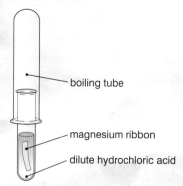

boiling tube

magnesium ribbon

dilute hydrochloric acid

b What happens if you apply a lighted splint to the boiling tube? What does this show?
c Name one metal that does not react with dilute acid.

3 a Name three fossil fuels.
b Explain what happens in the experiment below:

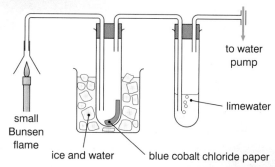

to water pump

limewater

small Bunsen flame

ice and water   blue cobalt chloride paper

c What is the chemical name for 'burning'?
d Copy and complete this word equation using key words from the list below:

methane + _____ → _____ _____ + _____

<< KEY WORDS >>

carbon dioxide   combustion   hydrogen   methane
oxygen   product   reactant   water

The particle model (7G)

1 Copy and complete a table like this one:

	Does it have its own fixed shape?	Is it easy to compress?	Does it spread out or flow easily?
solid			
liquid			
gas			

2
a Draw the particles in a solid, a liquid and a gas in three different boxes.
b Using the particle model, explain the changes you get as you heat a solid until it melts.
c Now explain the changes you get as you heat a liquid until it boils.

3 Which of these statements are true?
A: The particles in a solid do not move at all.
B: The particles in a gas zoom around randomly.
C: The particles in a solid expand as you heat the solid.
D: Particles escape from the surface of a liquid when it boils.

4 Describe the movement of the particles in each box in question 2.

5 **a** Explain why you can smell freshly baked bread from the oven in the kitchen throughout a house.
b What do we call the process in part **a**?

6 **a** What do we mean by 'gas pressure'?
b What causes gas pressure?
c Where do you think that gas pressure will be higher – on top of Mount Everest or at sea level?
Explain your answer.

7 A little water is boiled inside a thin metal can. The heating is stopped and a rubber bung is put in the top of the can to seal it.
After a few minutes the can starts to crumple up.
Explain these observations.

KEY WORDS

diffusion   evidence   gas pressure   liquid
model   particle   solid   theory   vibration

Solutions (7H)

1 Molly added a spoonful of sugar to a glass of water and stirred it with a spoon.
a Name the:
i) solvent   **ii)** solute
b Copy and complete using words from the key words box:

The sugar and water formed a
_____ . When no more sugar will dissolve we have a _____ solution. However, sand does not _____ in water. We say that sand is _____ in water.

c Here is a diagram showing the particles of a soluble red solid before it dissolves in water:
Draw a similar diagram to show the particles in the solution formed.

2 A food scientist was testing for a banned dye. Here is her chromatogram:

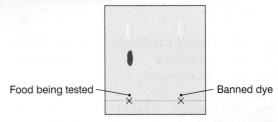

Food being tested   Banned dye

What does this chromatogram tell the scientist?

KEY WORDS

chromatography   dissolve   distillation
filtration   insoluble   saturated solution
soluble   solute   solution   solvent

Atoms and elements (8E)

1 Define the following words:
a atom
b molecule
c element
d compound.

2 Sort out these substances into elements and compounds:

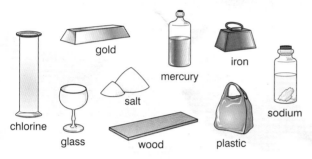

gold
iron
mercury
salt
sodium
chlorine
glass
wood
plastic

3 Give the chemical symbol for these elements:
a zinc
b nitrogen
c sulphur
d copper
e iron
f sodium

4 Look at this molecule of water:

Which statements are true:

A: Water is a chemical element.
B: Water is a compound.
C: The chemical formula of water is H_2O.
D: A molecule of water contains two oxygen atoms.
E: A molecule of water contains two hydrogen atoms.
F: Water is made up of two chemical elements.

5 Copy and complete these word equations:
a copper + oxygen → _____
b zinc + _____ → _____ sulphide
c potassium + bromine → _____
d _____ + _____ → sodium chloride

╾╼ **KEY WORDS** ╾╼

atom   compound   element   formula   molecule
Periodic Table   symbol

Compounds and mixtures (8F)

1

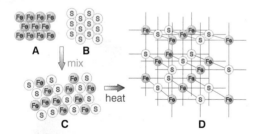

A   B
mix
C   D
heat

Explain which letter(s), A to D, shows:
a a mixture   b a compound
c an element.

2 The melting point of the element bromine is −7 °C. Its boiling point is 59 °C.
What will be the state of bromine at the following temperatures? (Solid, liquid or gas)
a 0 °C   b 100 °C   c 20 °C

3 Which statements about mixtures are true?
A: Mixtures have a fixed composition.
B: Mixtures can be separated by physical means, such as filtration or distillation.
C: Sodium chloride is a mixture.
D: Seawater is a mixture.
E: Air is a mixture of gases.

4 Use this table to answer the questions below:

Gases in the air	Formula of gas	Approximate proportions
nitrogen	N_2	78%
oxygen	O_2	21%
carbon dioxide	CO_2	about 0.04%
water vapour	H_2O	(varies)
argon (about 0.9%) and other noble gases	Ar, He, Ne, Kr, Xe, Ra	1%
various pollutants	e.g. SO_2 or NO_2 or CH_4	

a Which gases are made up of molecules of elements?
b Which gases exist as atoms?
c Which gases are compounds?
d Give the formula of the gas in air that is made up of molecules containing five atoms.

╾╼ **KEY WORDS** ╾╼

composition   compound   element   formula
impurity   mixture   proportion   pure   ratio

Rocks and weathering (8G)

1 Explain why sandstone is a porous rock but granite does not absorb water.

2 Which statement is true?
A: Granite is an element.
B: Granite is a mixture of elements.
C: Granite is a mixture of compounds.
D: Granite is a compound.

3 Explain what is happening in this diagram:

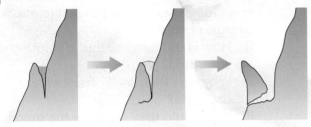

4 a Give four ways in which weathered rock is transported to another place in nature.
b Explain the difference between weathering and erosion.
c What type of rock fragments are usually deposited at the mouth of a river where it joins the sea?

5 Look at these layers of rock:

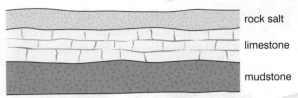

In which order were these rocks formed?

6 Explain how an evaporite rock, such as rock salt, was formed.

7 a Name two rocks that were formed from the remains of living things.
b What do we find in these types of rock?

The rock cycle (8H)

1 Look at these sediments being crushed together:

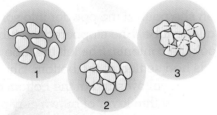

a What do we call this process?
b The pressure forces water out from between sediments and it can evaporate leaving behind any dissolved minerals. These minerals stick the grains of sediment together.
What do we call this process?

2 List three types of sedimentary rock.

3 a How are metamorphic rocks formed?
b Name three examples of metamorphic rock.

4 Explain how limestone can be changed into marble.

5 a What do we call molten rock found beneath the Earth's surface?
b What do we call molten rock that comes out of volcanoes?

6 Explain why granite and basalt are made up of different sized crystals.

7 Look at the rock cycle below:

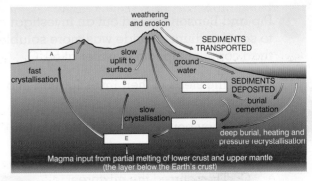

What are the missing labels?

1 Mike used a pH sensor to follow the reaction between sodium hydroxide solution and dilute hydrochloric acid. He added 2 cm³ of sodium hydroxide at a time and recorded the pH after each addition.

Here are his results:

Volume of sodium hydroxide solution added (cm³)	pH of the resulting mixture
0.0	5.0
2.0	5.0
4.0	5.0
6.0	5.5
8.0	6.0
10.0	7.0
12.0	8.0
14.0	8.5
16.0	9.0
18.0	9.0
20.0	9.0

a) Draw a graph of Mike's results. Make sure you label the axes, plot each point and draw a smooth curve. (4)

b) If Mike added another 2 cm³ of the sodium hydroxide solution, what is the pH likely to be? (1)

c) What do we call the type of chemical reaction observed in this experiment? (1)

2 Pip and Benson carried out an investigation to see if sodium chloride was more soluble than copper sulphate in water at 50 °C.

a) Put the steps in their investigation into the correct order.

    A: They pour 25 cm³ of water into two beakers.
    B: They recorded their results.
    C: They stirred the mixtures.
    D: They added sodium chloride to one beaker of water at 50 °C and copper sulphate to the other beaker of water at 50 °C.
    E: They heated the water to 50 °C. (1)

b) What piece of equipment did they use to make sure they had 25 cm³ of water in each beaker? (1)

c) They used 25 cm³ of water in each beaker, but what else did they do to make their investigation fair? (1)

d) Pip and Benson counted the number of spatulas of each solid needed to make a saturated solution.
    i) How could they tell that a solution was saturated? (1)
    ii) Why was their method not very accurate? (1)
    iii) How could they measure the amount of solid dissolved more accurately? (1)

e) Pip predicted that sodium chloride was more soluble in water at 50 °C than copper sulphate. They found that 10 spatulas of copper sulphate dissolved in their investigation. How many spatulas of sodium chloride might dissolve in a test that supported Pip's prediction?

    A: 2   **B:** 7   **C:** 10   **D:** 13 spatulas (1)

3 Reese hammered a piece of yellow sandstone into small pieces. She then crushed the bits into a powder.

Then she poured the powder into some water in a long tube. She stoppered the tube and shook the contents. She left the tube until the solids had settled at the bottom. She could see a layer of yellow sand at the bottom of the tube and a thin layer of brown clay on top of it.

a) What laboratory equipment did Reese use to grind up the bits of sandstone into powder? (1)

b) How could Reese tell from her results that yellow sandstone is a mixture of substances? (1)

c) Reese then filtered the mixture in the long tube. A clear, colourless liquid was collected.
    i) What could Reese do to see if any solids were dissolved in the colourless, clear liquid? (1)
    ii) How could she tell if there were any dissolved solids in the liquid? (1)

d) Sandstones are sedimentary rocks. Put these stages in the formation of a sedimentary rock into the correct order.

A: deposited
B: transported
C: weathered
D: compacted (1)

e) Sand contains silica. Its chemical name is silicon dioxide.
  i) The chemical symbol for silicon is Si. What is the formula of silicon dioxide? (1)
  ii) Is silica an element, a mixture or a compound? Explain your answer. (2)

 4 Methane can exist as a solid, a liquid or a gas depending on the temperature. The particles in the boxes represent methane molecules.

 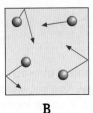 

    A   B   C

a) Which box contains:
  i) a solid
  ii) a liquid
  iii) a gas (1)

b) The melting point of methane is $-183\,°C$ and its boiling point is $-162\,°C$.
  i) What is the physical state of methane at $-170\,°C$? (1)
  ii) The formula of methane is CH_4. Which elements make up methane? (2)
  iii) When methane burns in plenty of oxygen, carbon dioxide is formed. Name the other product. (1)

Answers

1 a) Volume of sodium hydroxide on the x axis and pH of mixture on the y axis. (1)

Even scale chosen to fill most of the graph paper. (1)

11 points plotted correctly. (1)

Smooth curve of best fit drawn (not point to point) (1)

b) 9.0 (1)

c) Neutralisation (1)

2 a) A E D C B (1)

b) A measuring cylinder (1)

c) Used the same temperature/50 °C (1)

d) i) Some solid left undissolved. (1)
  ii) Each spatula measure might not have the same amount of solid on it. (1)
  iii) Measure the mass/use a balance (1)

e) D: 13 spatulas (1)

3 a) Mortar and pestle. (1)

b) She sees a yellow and a brown layer/it contains more than one substance which are not chemically combined/they can be separated by physical means or sieving. (1)

c) i) Heat the liquid/evaporate off the water. (1)
  ii) A deposit/solid/crystals /salt is left behind. (1)

d) C B A D (1)

e) i) SiO_2 (1)
  ii) A compound. (1)

It contains more than one type of atom chemically combined or bonded together. (1)

4 a) i) C   **ii)** A   **iii)** B (1)

b) i) Liquid (1)
  ii) Carbon (1)
  Hydrogen (1)
  iii) Water/steam (1)

911 Storing and using energy

If you lift a heavy object above your head, you have given it **potential energy**. You know that it has energy, because if you drop it and it lands on your toe – ouch! Another name for 'potential energy' is 'stored energy'.

Lots of things store potential energy. For example:

- a stretched rubber band
- a raised hammer
- a wound-up spring in a clock
- a cylinder of compressed air
- a spinning flywheel.

The heavier this container is and the higher it is lifted, the more potential energy it stores

Fossil fuels such as coal, oil and gas also store energy. We usually call this **chemical energy**. When the fuel is burned, its stored energy is released.

A tanker is waiting to load up with oil at this oil well. The energy of natural gas is being wasted as it is burned off.

A car runs on petrol, which is a store of chemical energy. The petrol burns and the car starts to move – it gains **kinetic energy**.

We use arrow diagrams (like the one shown here) to show energy changes like this. We say that the energy has been **transformed** from one form to another.

chemical energy (of petrol)　　　　kinetic energy (of car)

We use electricity a lot. It's a very convenient way of **transferring** energy from one place to another. The energy usually comes from burning gas or coal at the power station. Underground cables bring it to your home and you can plug in to make use of it.

We have lots of useful devices that use electricity and transform its energy into different forms.

◀ CHECKPOINT ▶

Electricity is a good way of **transferring/transforming** energy from place to place.

912 Battery power

Two or more **cells** connected together make a **battery**. (If you buy a 1.5 V 'battery', you are really buying a cell.)

You need cells to make a torch work. Inside a cell, there are chemical substances. When you switch on the torch, chemical reactions happen inside the cells and a current flows.

We say that a cell is a store of **chemical energy**. The bigger the battery, the more chemicals it contains, so it stores more energy and it will last longer.

When the current flows, it carries electrical energy to the different components in the circuit.

- bitumen
- insulating sleeve with top cap
- top washer
- manganese dioxide
- zinc can

The chemicals inside a cell are hazardous – they are corrosive and they could poison you. Don't open one up!

Voltage

The voltage is usually marked on a battery. It might say '1.5 V' or '3 V'. The V stands for **volts**. This tells you about the push it can provide to make a current flow in a circuit. The bigger the voltage, the bigger the push.

voltmeter

You can use a voltmeter to measure the voltage of a cell or battery. The illustration above shows you how.

> ### ◄ CHECKPOINT ►
>
> What voltage do you get from a battery made of four cells connected in series, if each one gives 1.5 V?
> _____

913 Generating electricity

Plug in and switch on. That's how we make use of electricity from the mains. But where does this electricity come from?

Most of our electricity comes from power stations. Inside, there is a generator which turns round and round, generating electricity. Usually, the generator is turned by steam; sometimes it's turned by flowing water or by the wind.

An old-fashioned bicycle dynamo works in the same way – as you pedal along, the back wheel turns the dynamo to provide electricity for the lights.

Most of our electricity comes from power stations where fossil fuels are burned, or where nuclear fuel (uranium) is used

The cost of electricity

We have to pay for the electricity we use. That's because the fuel for a power station costs money. The more we use, the more we pay.

Every house has an electricity meter to record how much electricity has been used. You can save money if you know how quickly different appliances use electricity. Look for the power rating, shown in watts (W) or kilowatts (kW).

- A 100 W lamp uses energy faster than a 60 W lamp.
- A 2 kW electric heater uses energy twice as fast as a 1 kW heater.

The longer you run an appliance for, and the greater its power rating, the more it will cost.

> ### ◄ CHECKPOINT ►
>
> How is electricity being generated in the photograph?
>
>

914 Measuring voltages

In a circuit, a cell or battery does two things:

- it provides the push to make the current flow

- it provides the energy to make the different components work.

To give a bigger voltage, connect two or more cells together in series. Then their voltages add up.

Using a voltmeter

When a current flows, there is a voltage across each component in the circuit. You can use a voltmeter to measure the voltage of the battery, and to measure the voltage *across* each component.

- Plug two leads into the voltmeter

- Connect one lead to one side of the component, and the other to the other side.

Sharing voltage

The voltage of the cell or battery is shared out between the different components around a series circuit. So, if you add up the voltages across all the components, they will add up to the voltage of the battery.

It's easy to tell the difference between an ammeter and a voltmeter:

- **Ammeter** – measures current – connected in series

- **Voltmeter** – measures voltage – connected in parallel.

The voltage of the battery provides the 'push' which is 'used up' by the components as the current flows through them

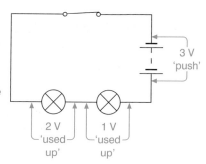

> **◀ CHECKPOINT ▶**
> The voltages around a **series/parallel** circuit add up to the voltage supplied by the battery.

915 Voltage and energy

The battery or power supply in a circuit provides the voltage to push current around. The voltage is shared out between the different components – we can say that they 'use up' the voltage.

Here's another way to think about this: the battery or power supply provides the energy, which is transformed by the different components.

The bigger the voltage, the faster the energy is being transformed.

- The voltage of a single cell is usually 1.5 V

- The mains voltage is 230 V

- The voltage of power lines can be as much as half a million volts!

Another name for voltage is **potential difference**.

Beware: very high voltages!

Thinking about current and energy

Current flows all the way around an electric circuit. (You can test this using an ammeter.)

As it flows, it carries energy with it. It collects energy from the battery or power supply, and shares it out among the different components.

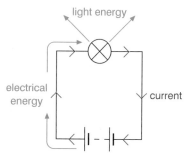

So, after the first component, there's just as much current flowing, but it's carrying less energy. After the last component, there's still the same current but it has no energy left – it must return to the battery to collect some more.

> **◀ CHECKPOINT ▶**
> Which quantity stays the same all round a circuit?
>
> _____
>
> Which quantity gets used up round the circuit?
>
> _____

Storing and using energy (911)

1 Join them up!

	a moving car
Potential energy	a raised hammer
	wood
Chemical energy	petrol
	a taut bowstring
Kinetic energy	food
	a falling stone

2 **a** In the first box below, draw an arrow diagram to represent what a telephone does: it transforms electrical energy to sound energy.

The second box shows an arrow diagram for a television set. It gives us light and sound (which we want), and it also gets hot (which is not really why we have a TV).

b In the third box, draw an arrow diagram for a mobile phone which can show photographs as well as letting you hear someone.

 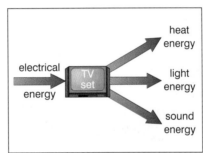 

3 What forms of energy are important here? Fill in the table.

What's going on	Form of energy
You stand on a pogo stick and its spring is squashed down	
You eat a large bowl of cereal and milk for breakfast	
You whiz to school at high speed on your bicycle	
You climb to the top of the ski slope, ready to slide downhill	
You wind up your grandmother's grandfather clock	

4 What energy transformations can you spot in the photograph?

Battery power (912)

1 Fill in the gaps:

   a A cell is a store of _____ energy.

   b An electric current carries _____ energy around a circuit.

   c When a torch bulb lights up, _____ energy has been transformed to _____ energy.

2 What instrument do we use to measure the voltage of a battery? _____

In the box on the right, draw the symbol for this instrument.

3 What is the unit of voltage? Give its name and symbol. _____

4 Look at the batteries and lamps in the photograph, and answer the following:

   a Which battery stores more energy? Explain how you know.

   _____

   b The smaller batteries last longer than the bigger ones. Why is this?

   _____

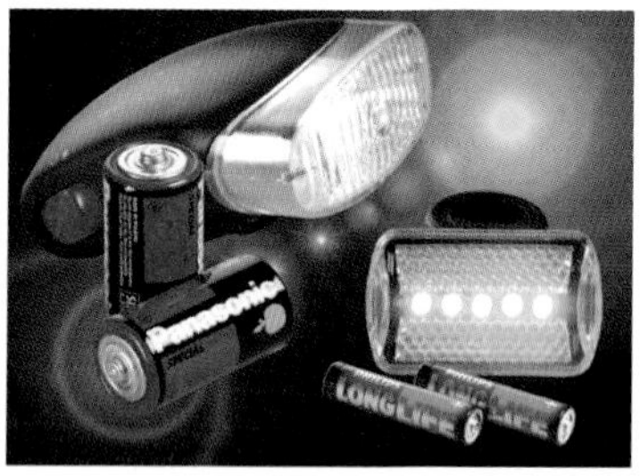

bigger batteries   **smaller batteries**

Generating electricity (913)

1 Fill in the gaps:

   a Most of our electricity comes from _____ _____.

   b The part which turns to produce electricity is called the _____.

   c The energy comes from burning _____ _____, or from a nuclear fuel such as _____.

2 Power rating tells us how fast an appliance uses electricity. What is its unit? Give the name and symbol.

_____ _____

Oh no, they've converted us to hydroelectricity!

3 Which uses energy faster? (Circle the correct answers.)

a 5 kW cooker **or** a 2 kW heater   one 100 W lamp **or** two 60 W lamps

4 Here are your electricity meter readings at the start of the year, and at the end of the year:

Jan 1st: 23020   Dec 31st: 24570

   a How many units of electricity have you used? _____

   b If each unit costs 8 p, what is the total cost? _____

5 Find out: what method of generating electricity doesn't make use of a rotating generator?

_____

Measuring voltages (914)

1 Fill in the gaps:

   a   A voltmeter measures the voltage _____ a component.

   b   A voltmeter must be connected in _____ with the component.

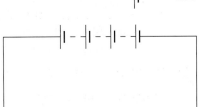

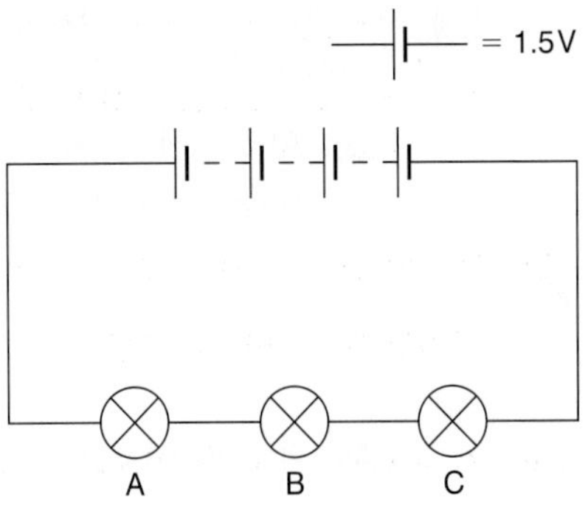

2 Look at the circuit shown on the right.

   a   What is the voltage provided by the battery of four cells? _____

   b   What is the voltage across lamp C? _____

3 Complete the circuit shown in box on the right, as follows:

   a   The empty circle represents a meter. Add the correct letter, to show what type of meter it is.

   b   What does it measure? _____

   c   Add a second meter to the circuit, to show how you would measure the voltage across the lamp.

   d   What would you expect the reading on this meter to be? _____

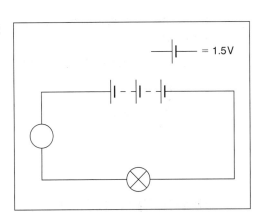

Voltage and energy (915)

1 Give another name for voltage. _____

2 In the box on the right:

   a   Underline all the components which can provide the voltage in a circuit.

   b   Circle the meters.

   c   Cross out all the components which use up the voltage.

voltmeter	cell	buzzer
lamp	ammeter	
power supply	resistor	
motor	battery	heater

3 Here are three ways we get electricity:

- from the mains
- from a power station
- from a battery

   a   Tick the one with the highest voltage.

   b   Put a cross by the one with the lowest voltage.

4 Find out what signs are used to warn of the dangers of the electricity supply.

Draw one or two electrical hazard symbols in the box.

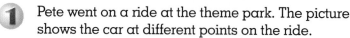

APPLY YOUR KNOWLEDGE

1 Pete went on a ride at the theme park. The picture shows the car at different points on the ride.

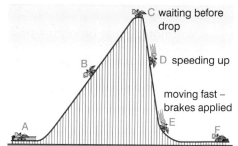

C waiting before drop

B

D speeding up

moving fast – brakes applied

E

A

F

a) At which point did the car have most potential energy? Explain how you know.

_____

_____ (2)

b) Name two points at which the car had no kinetic energy.

_____

_____ (2)

HINT

Think about how fast the car was moving.

c) At which point did the car have most kinetic energy? Explain how you know.

_____

_____

_____ (2)

2 Pip set up an electric circuit as shown.

a) Were the lamps connected in series or in parallel?

_____ (1)

2.0   2.5

b) How many volts did the battery supply?

_____ (1)

Pip wanted to check the voltage of the battery. The picture shows her circuit diagram.

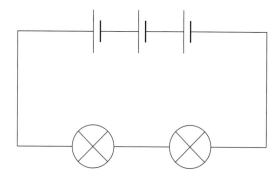

HINT

To decide on series or parallel, trace the path of the current round the circuit. Does it ever divide?

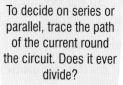

HINT

Remember that you need two extra leads for a voltmeter.

c) Add a voltmeter to the diagram, to show how Pip could measure the voltage of the battery. (1)

d) Pip also wanted to measure the current flowing around the circuit. On the diagram, mark with an **A** a point in the circuit where she should include an ammeter to do this. (1)

Before he went camping, Mike charged up his rechargeable torch.

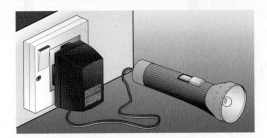

a) What energy change took place in the torch as the batteries were recharged? Circle the correct one.

chemical to electrical   electrical to light

electrical to light   electrical to chemical   (1)

b) What energy change took place when Mike used the torch in his tent at night? Circle the correct one.

electrical to chemical to light   electrical to potential to light

chemical to electrical to light   chemical to heat to light   (1)

First think about what is changing, then think about the energy change involved in this.

Gravity and space

9J1 Exploring the Moon

The Moon is much smaller than the Earth, so its gravity is much weaker.

It's the mass of the Moon that makes the difference. An object with a small mass has a weaker pull than an object with a bigger mass.

When astronauts went to the Moon, they had to learn to move around in conditions of low gravity. It was tricky, but they also had some fun. They could throw things much farther than on Earth, because everything weighed much less.

It's the pull of gravity that causes us to have **weight**.

● Mass and weight

● Because we are made of matter (stuff), we have **mass**. Mass is measured in kilograms (kg). It doesn't change if you go from one planet to another.

● Because the Earth has mass, it attracts things with its gravity. The pull of gravity is called weight. Because weight is a force, it is measured in newtons, N.

So mass is pulled on by gravity, and it also causes gravity.

The Earth's gravity pulls everything towards its centre

◄ CHECKPOINT ►

If you climbed Mt Everest, your **mass/weight** would be less because you would be further from the **centre/surface** of the Earth.

9J2 A trip into space

The launch of a spacecraft is an awesome sight. It takes a lot of energy to get a craft into space, or even to the Moon, so a lot of fuel must be burned.

You aren't in space until you are above the Earth's atmosphere. The atmosphere is a thin layer of gas around the Earth, a few hundreds of kilometres thick. It is held in place by the Earth's gravitational pull.

Because the Moon's gravity is much weaker than the Earth's, it has no atmosphere. (It may have had one once, but it has now all escaped.)

● Forces for flight

Even above the atmosphere, you are still pulled in by the Earth's gravity. The rocket provides the upward push needed to keep you moving away from the Earth.

If you switch off the rocket, you will fall back to Earth – splat! For a safe landing, you need to use rockets working in reverse, to slow you down. Alternatively, the friction of the atmosphere can slow a spacecraft, but you have to be careful – it can get very hot.

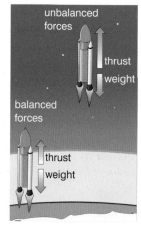

● If the forces on your craft are balanced, your speed will not change

● If the forces are unbalanced, you will speed up.

◄ CHECKPOINT ►

The Moon's gravity is weak because it has no atmosphere. True/False?

9J3 Exploring the solar system

Spacecraft from Earth have visited all nine planets of the solar system, so we know a lot more about them than if we could only use telescopes based on Earth. Cameras on board the craft take photographs and make other measurements which are then radioed back to Earth.

In the night sky, the planets look similar to the stars – bright specks of light. But they are different. The pattern of the stars remains the same from one year to the next, but the planets are seen to move against the background of the stars. That's how ancient people realised they are different, and why they named them after their gods – Jupiter, Venus and so on.

Planets are solid. Most have an atmosphere. Stars are giant balls of hot, glowing gas.

The night sky is full of stars which make up the patterns of the constellations. The more powerful your telescope, the more stars you will see.

The heliocentric solar system

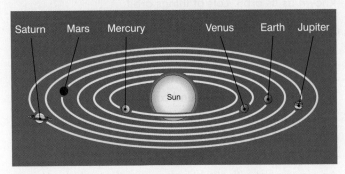

The six inner planets of the solar system

The Sun, planets and their moons make up the solar system. We can draw the system on a flat piece of paper because the planets travel in orbits which are nearly in the same plane.

Today, we know that the Sun is at the centre of the solar system, but this isn't obvious. From the Earth, it looks as though the Sun orbits us, rather than the other way round, because we see the Sun travel across the sky every day, and we don't feel the Earth moving.

However, over three centuries ago, astronomers proved that it is the Earth that orbits the Sun. This was the best way to explain the measurements they had made of the way that Mars moved across the sky.

Most people now believe today's picture of the solar system, with the Sun at the centre. It's called the **heliocentric model** of the solar system. (Heliocentric means 'Sun-centred'.) The old idea, with the Earth at the centre, was called the **geocentric model**.

> **◄ CHECKPOINT ►**
>
> The Moon orbits around the _____.
>
> The Earth orbits around the _____.

9J4 Staying in orbit

Each planet orbits around the Sun. It follows its own path, known as its orbit. Lots of things orbit:

- the Moon orbits the Earth
- washing in a washing machine, during the spin cycle
- the hands of a watch
- traffic on the M25 around London.

Each of these things needs an unbalanced force to make it go round in a circle. That's because, without an unbalanced force, things move in straight lines.

An experiment with a rubber bung on the end of a string shows this:

- whirl the bung round in a circle; you have to pull on the string to keep the bung in orbit
- let go of the bung; now there is no force pulling on the bung, and it flies off in a straight line.

It's the pull of the Sun's gravity that keeps the Earth in its orbit. The Sun's gravity pulls on all of the planets, even distant Pluto, so that they do not disappear off into space.

A **satellite** is the name we give to an object that orbits another.

> ◄ **CHECKPOINT** ►
>
> The Moon is held in its orbit around the Earth by the **pull/push** of the **Earth's/Sun's** gravity.

9J5 Satellites at work

There are hundreds of spacecraft in orbit around the Earth. An orbiting spacecraft is called an **artificial satellite**.

They have many different uses:

- they take photographs and make measurements to help weather forecasters
- they monitor the environment on Earth – to see if deserts are expanding, for example
- they help to send telephone messages around the world
- they broadcast satellite TV channels
- they are used by the military to spy on other countries
- they carry telescopes or other scientific instruments.

Once a satellite is in orbit, it doesn't need to use its rocket motors. It is travelling at a steady speed, so it doesn't need to burn any fuel.

> ◄ **CHECKPOINT** ►
>
> What force keeps an artificial satellite in orbit around the Earth? _____

This is Skylab, a scientific lab in space. You can see the solar panels it uses to generate the electricity needed to keep its instruments working.

Exploring the Moon (9J1)

1 What are the units for these quantities? Give the name and symbol.

Weight _____ _____   Mass _____ _____

2 Jupiter is a giant planet, with a mass much greater than the Earth's. If you went to Jupiter:

   a Your mass would stay the same / increase / decrease

   b Your weight would stay the same / increase / decrease.

3 Explain why we use an arrow to represent the weight of an object.

_____

4 The diagram shows someone's idea of how the Earth's gravity pulls on people at different places around the Earth.

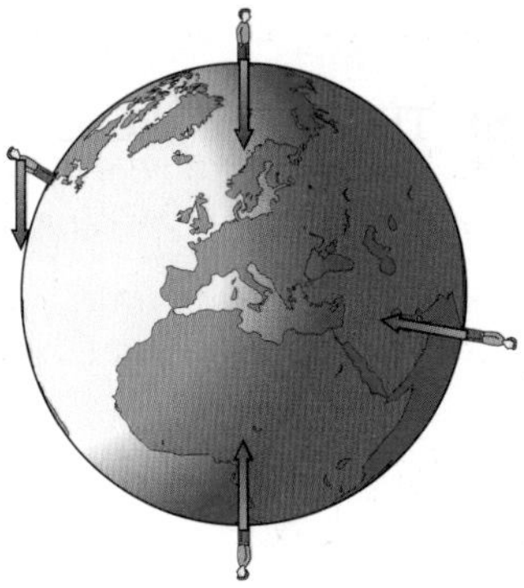

   a Put a tick next to any correct arrows.

   b Put a cross next to any incorrect arrows.

A trip into space (9J2)

1 Which force pulls downwards on a rocket as it is launched into space?

_____

_____

2 A rocket is leaving the launch pad.

   a In the space on the right draw the rocket.

   b Add arrows to represent the two forces acting on the rocket. (Which arrow should be longer?)

   c Are the forces on the rocket balanced or unbalanced? _____

   d How do you know? _____

3 Read the statements i–iv.

Copy the numbers i–iv into the boxes below in the correct order so that the sentences make sense.

   i) The Moon cannot hold on to its atmosphere.

   ii) The Moon's gravity is weak.

   iii) Astronauts must carry oxygen.

   iv) The mass of the Moon is smaller than the Earth's.

   ☐ **so** ☐ **so** ☐ **so** ☐

Exploring the solar system (9J3)

1 The photo on the right was taken from a spacecraft orbiting the Moon. List all the differences you can see between the Earth and the Moon.

____

____

____

____

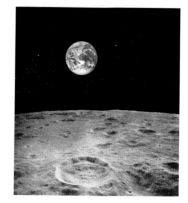

2 The picture shows the Sun and the inner planets of the solar system.

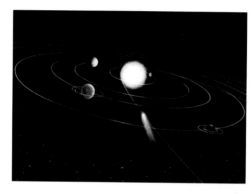

a Why is this model of the solar system called 'heliocentric'?

____

b Which planets can you see? ________ ________ ________ ________

c Circle the names of the planets which have moons.

d What is the other object with a long, glowing tail? ______________

3 The photograph shows the Eagle Nebula, a giant cloud of dust far off in space. New stars can be seen forming, and some of them probably have planets orbiting around them.

The photo was taken by the Hubble Space Telescope (HST).

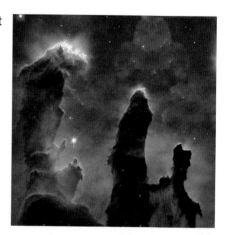

Do some research on space telescopes. Write a paragraph explaining:

a What is a space telescope?

b Why are they helpful to astronomers?

c How far are the most distant objects that a space telescope can see?

____

____

____

____

Staying in orbit (9J4)

1 Fill in the missing word. Choose from: rotating   orbiting   revolving   spinning.

If one object travels around another along a circular path, we say that it is _____ around it.

2 If an object follows a circular path, the forces on it must be **balanced/unbalanced**.

3 Name the Earth's natural satellite. _____

4 The diagrams on the right show the forces acting on four different objects. All of the objects are moving.

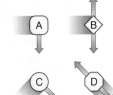

   a Which two are travelling at a steady speed in a straight line? _____ and _____

   b Which two could be travelling in a circle? _____ and _____

5 The picture shows a planet in orbit around the Sun. Add the following to the picture:

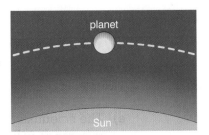

   a An arrow to show the force which keeps the planet in its orbit.

   b Next to the arrow, write the name of the force.

   c Using a different colour, draw a line to show how the planet would move if the force suddenly stopped pulling it.

Satellites at work (9J5)

1 Draw lines joining each type of satellite to the corresponding use.

Weather	transmitting telephone calls
Military	looking at distant galaxies
Communications	photographing clouds
Astronomy	studying the destruction of rainforests
Environmental	watching troop movements

2 A satellite travels around the Earth in a circular orbit once every 90 minutes. How are these things changing?

   a Its speed: increasing / decreasing / staying the same

   b Its potential energy: increasing / decreasing / staying the same

   c Its kinetic energy: increasing / decreasing / staying the same

   d The amount of fuel it carries: increasing / decreasing / staying the same

An astronaut in orbit around the Earth. He's moving at 8 km per second.

3 Find out about the work of astronauts in space. List some of the useful tasks they can perform.

_____

_____

_____

APPLY YOUR KNOWLEDGE

(SAT-style questions)

1 The picture shows two astronauts on the Moon. One is lifting up a moon rock.

a) Which arrow shows the correct direction of the force of the Moon's gravity on the rock? (1)

b) Which arrow shows the direction of the force of the astronaut's hand on the rock? (1)

The astronauts weigh their moon rocks using a newton meter and pack them into the spaceship. They climb aboard and set off for the return trip to Earth.

c) What happens to the pull of the Moon's gravity on the spaceship as it travels away?

_____ (1)

d) What happens to the pull of the Earth's gravity on the spaceship as it gets closer to Earth?

_____ (1)

e) The scientists weigh their moon rocks when they get back to Earth. What difference will they notice?

_____ (1)

f) Explain why the rocks have a different weight on Earth than on the Moon.

_____ (1)

2 The picture shows a spacecraft in orbit around the Earth. Its rocket motors are switched off.

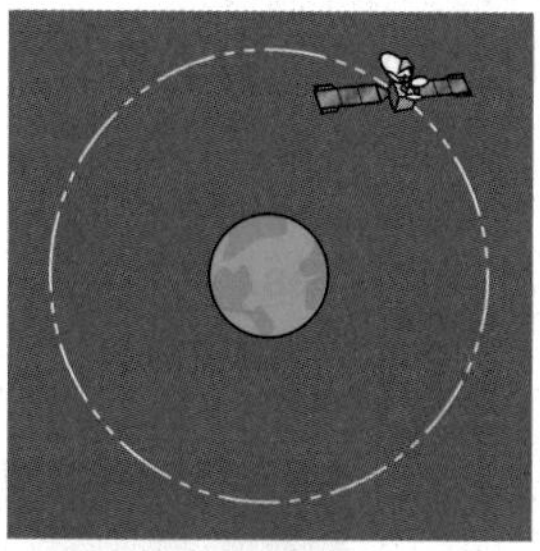

> **HINT**
> Remember that weight is the pull of gravity. What causes gravity?

> **HINT**
> Remember that gravity pulls things towards the centre of the Earth.

a) Add an arrow to the drawing to show the direction of the force of the Earth's gravity on the spacecraft. (1)

b) Describe how the spacecraft would move if the Earth's gravity was not pulling on it.

_____ (1)

c) Give **two** uses of spacecraft orbiting the Earth.

_____

_____

_____ (2)

> **HINT**
> Start each use with the word 'For ...'.

3 The photograph shows the Hubble Space Telescope.

Astronomers have used the space telescope to find out more about distant stars and galaxies.

a) Which method do these astronomers use to gain new information? (1)

Tick the correct one:

☐ They carry out experiments in their laboratories.

☐ They ask what other scientists think.

☐ They make observations of the environment.

☐ They find data on the internet.

HINT

You have to think about these astronomers, not what other astronomers might do.

b) Astronomers have benefited a lot by using spacecraft.

Suggest **one** way that spacecraft have helped astronomers gather new information about the solar system.

_____ (1)

c) Astronomers have found that there are many planets orbiting distant stars. What does this suggest about our knowledge of the solar system?

_____ (1)

d) Give **one** reason why scientists might reject an old idea and replace it with a new one.

_____ (1)

A photograph taken by a space telescope

Speeding up

9K1 Measuring speed

A car zooms past – how can you find its speed? You need to time the car as it moves a certain distance. Here's what you measure:

- the **distance travelled** by the car
- the **time taken** by the car.

Now you can calculate the car's average speed:

Average speed = distance travelled/time taken

Example

A car travels 120 m in 4 s. What is its average speed?

Distance travelled = 120 m
Time taken   = 4 s
Average speed   = 120 m/4 s = 30 m/s

(In science, we usually give speeds in metres per second.)

Average speed

We can only talk about the car's average speed, because we don't know if it is going at a steady speed. It might be slowing down, or speeding up. (If you looked at its speedometer, you'd be able to tell.)

Precision

When athletes run a sprint race, their times must be measured with great **precision**. For example, someone might win the 200 m sprint in 19.96 s. This means that their time has been measured with great precision, to the nearest 0.01 s.

This is not the same as saying that it is measured very **accurately**. If the timing was accurate, it would mean that it gave the correct time that the runner had taken.

◀ **CHECKPOINT** ▶

Give the scientific unit of speed. _____

9K2 Speeding up

To get a more precise measurement of speed in the lab, you can use **light gates**.

- When the car passes through the first gate, it starts the timer
- When it passes through the second gate, it stops the timer. Now you know the time taken
- You have to measure the distance between the gates to find out the distance travelled
- Now you can calculate the car's average speed.

If the light gates are connected to a computer, it can calculate the speed for you.

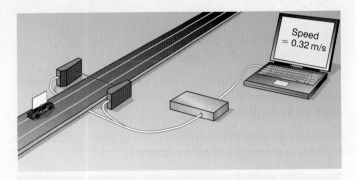

You can find a car's speed using a single light gate.

Stick a card, 5 cm long, on top of the car.

When the front edge of the card breaks the light beam, it starts the timer. When the back edge of the card passes through the beam, the timer stops. Now you know how long it took the car to travel 5 cm, and you can work out its speed.

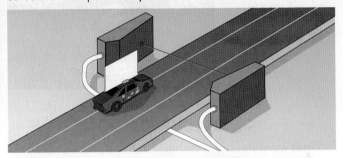

A light gate sends a beam of light across the track of the car; when the car breaks the beam, it sends a signal to the timer to start or stop

◀ CHECKPOINT ▶

A car travels 5 cm in 0.20 s. Circle its correct speed:

   1 cm/s   5 cm/s   25 cm/s   100 cm/s

9K3 Changing speed, changing direction

If you want to get off to a good start in a race, you need a big force to push you forwards. Then you will speed up or **accelerate** well, and you may get ahead of the rest.

The force that makes you start moving must be unbalanced. **Balanced forces** cancel each other out, so that they have no effect. If the forces on you are balanced you will:

- stay still, or
- move in a straight line at a steady speed.

There is little friction to slow down an ice skater. That means that they can easily travel in a straight line. It's more difficult when they want to travel in a curve – they have to push sideways on the slippery ice.

An **unbalanced force** can start you moving. There are other things that unbalanced forces can do:

- they can speed you up when you are already moving
- they can slow you down
- they can make you change direction.

If you are moving straight ahead, a force pushing from the right will make you curve off to the left.

🔵 Working out forces

The forces on an object must be balanced if it is stationary, or moving at a steady speed in a straight line.

The forces on an object must be unbalanced if its speed is changing, or if its direction is changing.

◀ CHECKPOINT ▶

What is the scientific word meaning 'to speed up' or 'to get faster'?

_____

9K4 It's a drag

Friction is a force which resists motion.

If one surface tries to slide over another, they rub together, and this causes **friction**.

If an object moves through water, it has to push the water aside. There is also friction with the water, and this all adds up to the force of **drag**.

An object moving through the air has to push air aside, and this is the origin of **air resistance**.

Jet skis are designed to ride up out of the water, because there's less resistance in air than in water. They have a streamlined shape, too.

Any vehicle moving along the road experiences air resistance, tending to slow it down. The driver has to press harder on the accelerator, to increase the forward force of the engine. As a consequence, the engine burns fuel faster, and that costs money.

At high speeds, air resistance is much greater, and so the cost of driving increases as you go faster. It would be much better if we lived on the Moon, where there is no air resistance!

◄ CHECKPOINT ►

What word describes a shape that is designed to reduce air resistance or drag?

_____

9K5 Going up, coming down

If you dive to the bottom of the swimming pool, you will soon come back to the surface. The force of **upthrust** is greater than your weight, and it pushes you back up. Divers who want to stay underwater have to weight themselves down with bits of lead, so that their weight is greater than the upthrust.

It's the same for hot air balloons. The surrounding air pushes up on them, so that they float gently upwards. The height can be controlled by letting hot air out of the balloon.

A parachutist makes use of air resistance.

● When they jump out of the aircraft, there is little resistance. Their weight pulls them downwards

● As they go faster, air resistance increases, but they keep falling faster

● Eventually, they are going so fast that there is enough air resistance to balance their weight. They fall at a steady speed

● When they open their parachute, there is much more air resistance. This slows them down to a new steady speed

● Bump! They land on the ground.

The graph shows the story of a parachute jump.

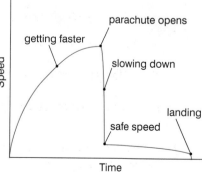

◄ CHECKPOINT ►

Air resistance **increases/decreases** as you go faster.

Measuring speed (9K1)

1 Put the following units in the correct columns in the table.

~~m~~   ~~s~~   ~~km~~   ~~seconds~~   ~~mph~~   ~~km/s~~   ~~hours~~   ~~miles~~   ~~metres~~   ~~m/s~~

Distance travelled	Time taken	Speed
metres ✓ km ✓ s ✓ miles ✓ m ✓	hours ✓ seconds ✓ s ✓	mph ✓ miles ✗ m/s ✓ km/s ✓

2 The Scientifica crew had a competition to see whose average speed was the greatest.

Benson said, 'I'm best at sprinting, so I'll run 100 m.'

Molly said, 'I'm best at cross-country.' She ran 5000 m.

In the end, they all ran different distances. Complete the table, and put a star next to the name of the person who was fastest.

Name	Distance run (m)	Time taken (s)	Average speed (m/s)
Pip	1000	125	8
Reese	120	14	8·6
Mike	200	32	6·25
Molly	5000	724	6·91
Benson	100	12.5	8
✱ Pete	400	24	16·67

Do you think this was a fair test of their running ability? Explain your answer. _i don't_ _think this is a fair test because they all ran at different speeds and distances._

3 A sheet of A4 paper is 297 mm long. Pip, Pete and Molly measured a sheet. Here are their measurements:

Pip: 'It's 29 cm long.'

Pete: 'It's 29.50 cm long.'

Molly: 'It's 29.7 cm long.'

a   Whose answer was the most accurate? ___Molly___

b   Whose answer was the least precise? ___Pip___

4 A supersonic aircraft travelled 12 000 km in 6 h. Work out its speed:

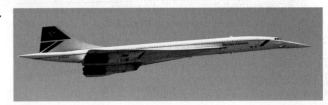

a   in km/h ___2000___

b   in m/s ___2,000,000___
___556___

Speeding up (9K2)

1 In toboggan races, each rider is timed. One light gate detects when they cross the starting line, and another when they cross the finishing line.

A tobogganist took 30 s to travel 900 m.
Calculate his average speed.

Average speed = distance travelled/time taken = _____ = _____ m/s

2 A car with a card attached, 6 cm long, took 0.40 s to pass through a light gate.

What was its average speed?

Average speed = _____

3 A local council sometimes surveys the traffic using a road. They place two strip detectors across the road; as a car passes over, each strip sends a signal to the recorder.

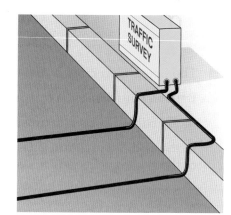

a How could this system be used to tell which direction the car was travelling in?

_____

b How could it work out how fast the car was travelling?

_____

Changing speed, changing direction (9K3)

1 The two photos, A and B, show runners at the start of a race.

a In which photo are the forces on the runners *balanced*? Photo ___. Explain how you can tell.

_____

b In which photo are the forces on the runners *unbalanced*? Photo ___. Explain how you can tell.

_____

A    B

2 Complete the table. In the second column, say whether the forces are balanced or unbalanced. In the first column, mark the words which tell you that the forces are balanced or unbalanced.

Situation	Balanced or unbalanced forces?
A train is travelling along a straight track at a steady speed.	
A car is accelerating as it travels along the motorway.	
A cyclist speeds around a curved track.	
A cat sits on a mat.	

It's a drag (9K4)

1 Name the forces that try to slow things down:

a When two surfaces rub together: _____

b When an aircraft flies through the air: _____

c When a shark swims through the sea: _____

2 Use what you know about air resistance to explain why racing cyclists travel one behind the other, as shown in the photograph.

_____

_____

3 Look at the photograph of the racing car.

a Explain why it is important to have a lot of friction between its tyres and the road.

_____

b Explain how the car has been designed to reduce the effects of air resistance.

_____

Going up, coming down (9K5)

1 Unscramble the names of these forces:

gard   prut shut   rise stair cane   inforcit   the wig

drag ✓   upthrust ✓   air resistance ✓   friction ✓   weight ✓

2 You may have been on a theme park ride like this:

- You are in a car which starts to run down a steep slope.
- You go faster and faster.
- At the bottom, the track levels off and the car slows to a halt.

a What force makes you go faster and faster? ___weight ✓___

b What force slows you down? ___friction ✓___

c Complete the graph on the right to represent your journey.

3 Free-fall parachutists don't open their parachutes until the last minute.

They can control their speed by changing the way they fall.

In the space on the right, draw sketches to show how you would arrange your body:

a to fall as fast as possible

b to slow yourself down.

12/12

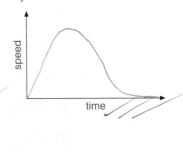

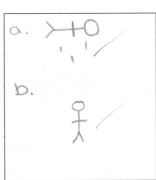

1 The picture shows Reese pushing a heavy box.

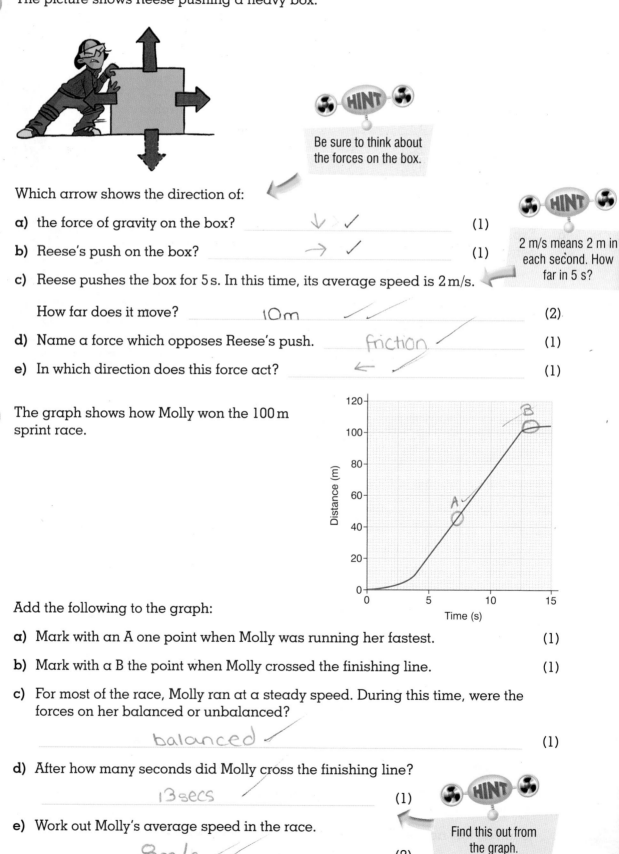

HINT

Be sure to think about the forces on the box.

Which arrow shows the direction of:

a) the force of gravity on the box? ___↓ ✓___ (1)

b) Reese's push on the box? ___→ ✓___ (1)

HINT

2 m/s means 2 m in each second. How far in 5 s?

c) Reese pushes the box for 5 s. In this time, its average speed is 2 m/s.

How far does it move? ___10m___ (2)

d) Name a force which opposes Reese's push. ___friction___ (1)

e) In which direction does this force act? ___←___ (1)

2 The graph shows how Molly won the 100 m sprint race.

Add the following to the graph:

a) Mark with an A one point when Molly was running her fastest. (1)

b) Mark with a B the point when Molly crossed the finishing line. (1)

c) For most of the race, Molly ran at a steady speed. During this time, were the forces on her balanced or unbalanced?

___balanced___ (1)

d) After how many seconds did Molly cross the finishing line?

___13secs___ (1)

HINT

Find this out from the graph.

e) Work out Molly's average speed in the race.

___8m/s___ (2)

3 Mike's little brother had a toy car. Mike made it run along the floor.

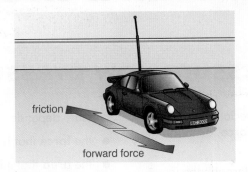

friction

forward force

a When the car was speeding up:

Friction was greater than the forward force.

(Friction was less than the forward force.)

Friction and the forward force were equal.

Circle the correct sentence. (1)

HINT

Start by deciding if the forces are balanced or unbalanced.

b When the car was travelling at a steady speed:

Friction was greater than the forward force.

Friction was less than the forward force.

(Friction and the forward force were equal.)

Circle the correct sentence. (1)

26/26

9L Pressure and moments

9L1 High pressure, low pressure

In science, the word **pressure** has a special meaning. It doesn't mean quite the same thing as **force**, although the two words are closely related.

Look at the picture. The boy has fallen through the ice. His weight was more than the ice could support. How can he be rescued?

If someone else tried to walk on the ice, they would find that it is too thin. They would fall in, too.

The answer is to lay the ladder on the ice, and then crawl along the ladder towards the boy. Then the rescuer's weight will be spread out over a bigger area, and the ice won't crack.

Please use your scientific knowledge to save me!

Pressure depends on two things:

● the **force** that is pressing downwards

● the **area** that it is pressing on.

To reduce the pressure, spread the force over a bigger area.

We often want a high pressure. For example, if you want to make a hole in a piece of leather, you should use a needle with a sharp point. The area of the point is small, so that there is a high pressure when you push on the needle.

So pressure tells you about how concentrated a force is; if a large force is concentrated on a small area, the pressure will be high.

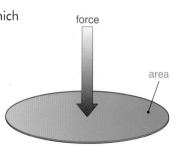

⫷ CHECKPOINT ⫸

Cross out the incorrect words; do this twice to make two different sentences.
● A **big/small** force pressing on a **large/small** area gives a **high/low** pressure.
● A **big/small** force pressing on a **large/small** area gives a **high/low** pressure.

9L2 Stiletto sums

◉ Calculating pressure

To calculate the pressure which a force exerts, we use this equation:

    Pressure = force/area

The force must be given in newtons (N), and the area in cm^2 or m^2. Then the pressure will be in N/m^2 or N/cm^2.

force

area

Here's an example of how to calculate pressure:

A person weighing 800 N lies on the ground. The area of contact with the ground is 0.4 m^2. What is the pressure on the ground?

    Pressure = force/area = 800 N/0.4 m^2 = 2000 N/m^2

If the same person stands up, the pressure will be greater because their weight is concentrated on a smaller area.

    Area of shoes = 0.004 m^2

    Pressure = 800 N/0.004 m^2 = 200 000 N/m^2

Stiletto heels have an even smaller area, so the pressure would be greater still.

◉ Pressure in liquids and gases

There is a famous story of a Dutch boy who saved his country by sticking his finger in a hole in the dyke. If he had pulled his finger out, water would have squirted out and drowned the land. The water behind the dyke was pushing on the dyke with great pressure. You rely on water pressure to make water come out of the taps or shower when you have a wash.

Gases can exert pressure, too. The air around you is always pressing on you.

You can understand about pressure in liquids and gases if you think about the particles they are made of. They are constantly moving around, bumping into the walls of their container. Each bump gives a small push on the container, and all of these tiny pushes add up to give the pressure.

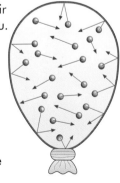

⫷ CHECKPOINT ⫸

Which are units of pressure? Cross out the incorrect ones.
N/m^2 N   m^2/N   N/cm^2

9L3 Hydraulics

Because liquids and gases can exert pressure, we can use them to transfer the effect of a force from place to place. Hydraulic digging machines and tipper trucks make use of this.

Oil or water is pushed into a cylinder, and this makes a piston move. A small force pushing on a narrow cylinder can become a big force pushing on a large piston. This is the basis of hydraulic machines.

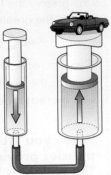

A small force can be magnified using a hydraulic system

Atmospheric pressure

If you dive down into the ocean, the pressure becomes greater. The water presses down on you.

In the same way, the atmosphere presses on you. We live at the bottom of the atmosphere. If you climb a mountain, you are climbing up through the atmosphere and the pressure gets less.

Atmospheric pressure is about 100 000 N/m². This is the same as 100 000 pascals (Pa).

It is easy to **compress** a gas such as air, because there is a lot of empty space between its particles. Liquids can be compressed a bit, because they have a bit of empty space. Solids are hardest to compress.

A pneumatic machine makes use of compressed air instead of oil or water.

> **◄ CHECKPOINT ►**
> Give another name for 1 pascal (1 Pa).
> _____

9L4 Levers everywhere

A gardener might use a lever to lift a heavy paving slab. That's just one use of a lever. There are many others which we use in everyday life, although we may not realise it: wheelbarrows, a pair of scissors (that's two levers working together) . . . If you look at the photo of the digger (above), you will see that it uses levers, too.

A lever is a device which enables us to do something more easily. We apply a force, and the lever makes the force do something useful.

Look out for the **pivot** of any lever. That's the point that doesn't move as the lever tips.

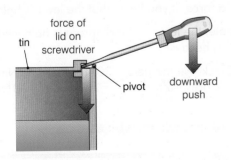

Levers in the body

Your muscles and bones work together to act as levers. A muscle pulls on a bone to make it move. A *different* muscle pulls on the bone to make it go the other way. This is because a muscle can only pull, it can't push. Two muscles that work in opposite directions like this are called an **antagonistic pair**.

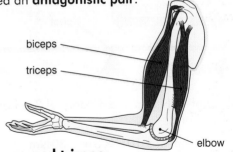

The biceps and triceps muscles of the upper arm form an antagonistic pair

> **◄ CHECKPOINT ►**
> Which joint of the body forms the pivot for the biceps and triceps muscles?
> _____

This cyclist makes good use of his leg-levers: one leg is pushing down while the other is pulling up

9L5 Getting balanced

Levers are useful because they can allow us to lift heavier loads than we could manage without a lever. Think about a wheelbarrow.

- The **pivot** is the wheel
- The **load** is the weight of the stuff in the barrow
- The **effort** is the force you provide to raise the end of the barrow.

You could lift a load of 100 N using a force of just 50 N. The effort is less than the load. This is because the load is close to the pivot, and the effort is farther from the pivot.

When a force is farther from the pivot, it has a greater **turning effect**. If it is twice as far from the pivot, it will have twice the turning effect.

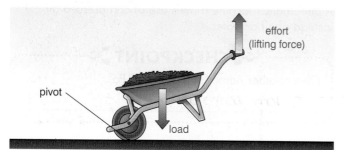

effort
(lifting force)

pivot

load

A see-saw is a type of lever, with the pivot at its middle. It can balance in different ways:

- Two people of equal weight can sit at opposite ends.
- Two light people can balance one heavy one.
- A light person at one end can balance a heavier person who is sitting closer to the pivot.

So the greater the force and the greater its distance from the pivot, the greater its turning effect.

---- **CHECKPOINT** ----

If you increase a force and move it towards the pivot, its turning effect will: **increase/decrease/can't say.**

When you lift a weight, your muscles provide the effort. The weight is the load. Your elbow is the pivot.

9L6 A matter of moments

The 'turning effect' of force has a special name: it is called the **moment** of the force. The greater the force and the further it acts from the pivot, the greater is its moment.

moment of force = force × distance from pivot

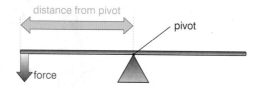

distance from pivot

pivot

force

So, if a force of 10 N acts at a distance of 3 m from a pivot, its moment is:

moment = 10 N × 3 m = 30 Nm

The unit of moment is the newton-metre (Nm).

It is important to work out whether a force is turning a lever clockwise or anticlockwise. Look at the diagram. Remember that the pivot is fixed; the force is trying to turn the lever anticlockwise, because it is pulling downwards on the left.

● Principle of moments

If a lever (such as a see-saw) is balanced, it means that it isn't turning clockwise or anticlockwise. The moments of the forces which are acting on it are cancelling each other out. So, if a lever is balanced:

$$\text{Sum of clockwise moments} = \text{Sum of anticlockwise moments}$$

This is called the **principle of moments**. You can use it to work out if a lever is balanced. You can also use it to work out a force, if you know that the lever is balanced.

For example, the lever in the diagram is balanced. We can say this because:

Clockwise moment = 30 N × 2 m = 60 Nm
Anticlockwise moment = 20 N × 3 m = 60 Nm

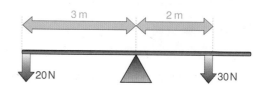

3 m   2 m

20 N   30 N

---- **CHECKPOINT** ----

If the 20 N force was reduced to 10 N, would the lever turn clockwise or anticlockwise? _____

What force could then replace the 30 N force to re-balance the lever? _____

High pressure, low pressure (9 L 1)

1 Which gives the biggest pressure? Put a tick in the correct box.

Which gives the smallest pressure? Put a cross in the correct box.

a big force pressing on a small area ✓ a   a big force pressing on a big area ☐

a small force pressing on a big area ✓ b   a small force pressing on a small area ☐

2 If you want to walk on soft snow, you may wear snowshoes. (They look like tennis rackets, strapped to your feet.) Explain why this helps you to walk on the snow.
Use the word *pressure* in your answer.

_____

_____

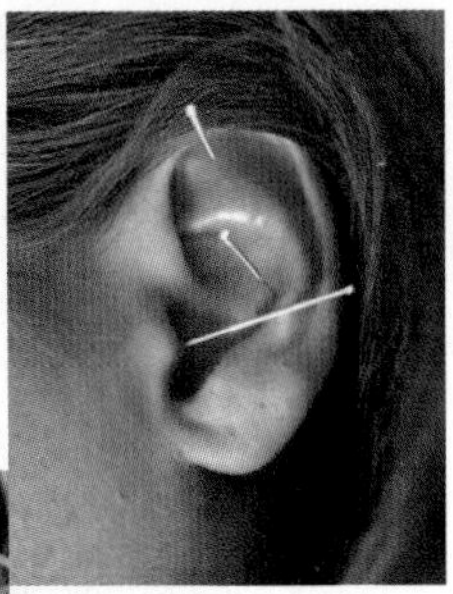

3 If you have acupuncture, needles are used to puncture your skin. Explain why it is best if they use sharp needles, rather than blunt ones. Use the word *pressure* in your answer.

high pressure and _____

_____

4 Why does the African Jacana have big feet?

to reduce the pressure on
the lily pads

The African Jacana can walk on water (almost)

Stiletto sums (9 L 2)

1 Complete the table to show the units of pressure.

Force	Area	Pressure
N	m²	N/m²
N	cm²	N/cm²

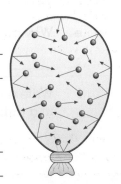

2 A force of 100 N presses on an area of 5 m². Complete the calculation:

pressure = force / area = _____ / _____ = _____ N/m²

3 Calculate the pressure caused by a force of 500 N pressing on an area of 12.5 cm².

Pressure = 500/12.5 = 40 N/cm² _____

4 The air presses on everything it touches with a pressure of 100 000 N/m².

a What force does it exert on a window of area 1 m²? _____

b What force does it exert on a large window of area 4 m²? _____

5 Explain how the particles of air inside a balloon keep it inflated.
(The picture may help.)

_____

_____

Hydraulics (9L3)

1 Look at the terms in the box.

    a   Underline all of the units.

    b   Put rings round the three terms which are the same
         as each other.

> Pa   N   N/cm^2
> pascal   pressure   N/m^2
> force/area   hydraulic

2 Give one word which means the same as: 'squash into a smaller volume'. _pressure_

3 Give one word which describes a system which uses a liquid to transfer and magnify a force.

_____

4 Look at the drawings which show the particles of a solid, a liquid and a gas.

    a   Label the drawings *solid*, *liquid*, *gas*.

    b   Use the drawings to explain why a gas can be
         compressed more easily than a solid or a liquid.

    solid   _liquid_   _gas_

_____

_____

_____

Levers everywhere (9L4)

1 A lever can allow us to use a small force to move a larger force. Give an example of this.

_____

2 The muscles and bones of our limbs work like levers.

    a   The biceps muscle is one of a pair of antagonistic muscles in our arm.
         Name the other. _____

    b   What is the pivot for this pair? _____

    c   Name the two muscles of an antagonistic pair in our leg. _____

    d   What is the pivot for this pair? _____

    e   Why are pairs of muscles like this described as 'antagonistic'? _____

_____

    f   Write a sentence using the word 'antagonistic'
         in a different sense (nothing to do with muscles).

quadriceps

biceps

hamstrings

triceps

elbow   knee

_____

_____

Getting balanced (9L5)

1 Which of the following will make a force have a greater turning effect?
Use ticks and crosses to show your answers.

Increasing the force   Increasing its distance from the pivot

Decreasing the force   Decreasing its distance from the pivot

2 In the picture, what will happen if Benson moves closer to the pivot?

_____

3 In the space on the right, draw a sketch to show how a small person can balance a heavier one on a see-saw.

4 When you bite on something, your muscles pull your jaws together – see the photo.

a   What force is the effort here? _____

b   What force is the load here? _____

c   Feel your cheeks, below your ears, as you make your jaws bite together. Where is the pivot? _____

A matter of moments (9L6)

1 What is unit of the moment of a force? _____ newtons   NM ✓

2 Calculate the moment of force of 40 N acting 0.5 m from a pivot.

force × distance

= 40 × 0.5 = 20 NM ✓✓

3 Pip carried out an experiment with weights on a ruler, suspended at its midpoint.

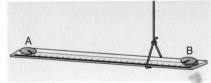

The table shows some of her results. In the last column, write 'B' if the ruler was balanced, and 'U' if it was unbalanced.

Weight A (N)	Distance from pivot (m)	Weight B (N)	Distance from pivot (m)	Balanced?
10	0.4	10	0.4	B
20	0.5	25	0.4	B
12	0.25	8	0.5	U
30	0.5	50	0.4	U

4 Pip then did some more balancing experiments. The table shows her results; fill in the blank spaces to show the forces and distances needed to balance the ruler.

Weight A (N)	Distance from pivot (m)	Weight B (N)	Distance from pivot (m)
10	0.5	20	0.25
20	0.4	16	
	0.3	6	0.4
10		10	0.4

1   The table shows the weights of three of the Scientifica crew.

Molly	600 N
Benson	650 N
Reese	570 N

a) Molly sat on one end of a see-saw.
    There was no-one on the other end.
    In which direction did Molly's end move?

    _____down_____ (1)

HINT

When looking at a lever, always think about whether each force is turning it clockwise or anticlockwise about the pivot.

b) Then Benson sat on the opposite end.
    He sat at the same distance from the pivot as Molly. How would the see-saw move?

    both be eqial. mollys one goes up(1)

c) Reese took Benson's place on the see-saw.
    She wanted the see-saw to be balanced.
    Who should sit closer to the pivot, Molly or Reese?

    _____Reese_____ (1)

HINT

Think about how their weights compare.

2   Pete picked up a heavy metal block. It weighed 30 N.
    It covered 20 cm² of Pete's palm.

HINT

Remember that a **big** force on a **small** area gives a **big** pressure. This will help you to remember how to calculate pressure.

a) What was the pressure of the weight on Pete's hand?
    Give the unit.

    _____30×20 = 500_____ (2)

The diagram shows the muscles in Pete's arm.

b) Which muscle must contract to lift the metal block?

    _____triceps_____ (1)

c) What must the other muscle do?

    _____ (1)

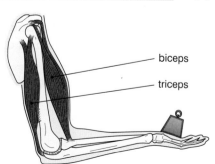

biceps

triceps

3 The diagram shows a barrier which can be raised using the metal cable.

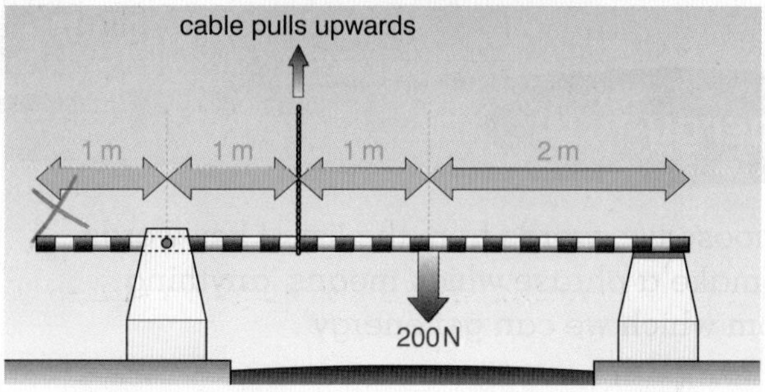

cable pulls upwards

1 m   1 m   1 m   2 m

200 N

a) Calculate the moment of the barrier's weight about the pivot. Give the unit.

1m _____ (2)

b) What force must the cable provide to raise the barrier?

_____

_____ (1)

HINT

The barrier is balanced, so you can calculate the force using the principle of moments.

c) Mike suggested that it would be easier to raise the barrier if a heavy weight was added to one end of the barrier. Mark the diagram to show where you would add a counterweight.

_____ (1)

d) Which would be heavier, the barrier or the counterweight? Explain your answer.

_____

_____ (2)

HINT

Think about which is closer to the pivot.

Energy (7I)

1 Choose two words from the list of key words to make a phrase which means 'anything from which we can get energy'.

2 Give one word which means 'a material we burn to release energy'.

3 Name three fossil fuels. What name do we give to a fuel which is made from material which has grown very recently?

4 Which two words in the list of key words are units of energy? Write the scientific unit of energy first.

5 We can generate electricity using the energy of the wind. Why do we describe the wind as a **renewable** source of energy?

6 In South Africa, petrol is made from coal. Is this a renewable or non-renewable resource? Explain your answer.

7 a Vegetarians eat nothing but plant material. Draw a diagram to explain how the energy used by vegetarians comes originally from the Sun.
  b Other people eat meat and other animal products. Add these people to your diagram.

8 The picture shows a simple experiment to compare the energy stored in different foods.

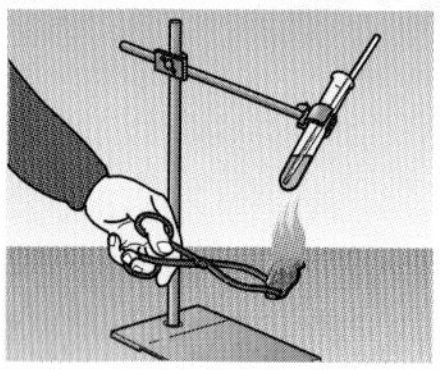

  a How would you decide which food contained the most energy?
  b How would you make this as fair a test as possible?

> ◀ **KEY WORDS** ▶
>
> biomass calorie conserve energy environment
> fossil fuel kilojoule renewable resource

Electrical circuits (7J)

1 From the list of key words, choose words which mean:
  a a device which prevents electric current from becoming too great
  b the tiny particles which move in wires when a current flows
  c two different ways of connecting components in a circuit.

2 a What is measured using an ammeter?
  b What are the units of this quantity?

3 Copy the diagram below.

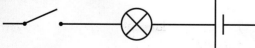

  a Add a wire to complete the circuit.
  b Label all of the components.
  c What will happen when the switch is closed?

4 We say that a fuse 'blows' when the current through it is too great. What actually happens to the wire inside the fuse? How does this protect the circuit?

5 Look at the picture of the circuit.
  a Are the components connected in series or in parallel?
  b Draw a diagram to represent this circuit. Include an ammeter to measure the current from the cell.
  c How does the electric current flow round this circuit? (You may want to add arrows to your diagram to show your answer.)

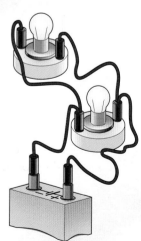

> ◀ **KEY WORDS** ▶
>
> ammeter battery cell current electrons
> energy fuse insulation model parallel
> resistance series voltage

Forces (7K)

1 Look at the list of key words. Which ones are the names of forces?

2 Give the units for each of the following quantities:

force   mass   density   weight   speed

3 Copy and complete:

If you go to the Moon, your _____ will decrease because the Moon's _____ is weaker than the Earth's. However, your _____ will stay the same, because you are made of the same amount of _____ .

4 Look at the drawing of the man sitting on the chair. Are the forces on him balanced? How can you tell?

downward pull of weight

upward push of chair

5 Look at the photo of the racing car. Explain how its design helps it to go fast. Use as many scientific words as possible in your answer.

6 Calculate the density of a rock:

mass of rock $= 1800\,\text{kg}$
volume of rock $= 1.2\,\text{m}^3$

7 If you put a piece of wood into water:
  a What force pulls on it, downwards?
  b What force pushes on it, upwards?
  c If the wood floats, what can you say about these forces?

KEY WORDS

air resistance   balanced   density   drag   forcemeter
friction   gravity   kilogram (kg)   lubricant   mass
newton (N)   speed   upthrust   weight

The solar system – and beyond (7L)

1 Copy the table. Choose the correct words from the key words list to fill the first column.

Key word	Definition
	turn about an axis
	travel along a path around another object
	an object which travels around another
	the layer of gas around a planet

2 Which three objects in the list of key words are satellites of the Sun?

3 **a** Draw a diagram to show the positions of the Earth, Moon and Sun when there is an eclipse of the Moon.
  b Show the Earth's shadow.
  c Mark with a tick a point on the Earth's surface from which the eclipse will be visible.
  d Mark with a cross a point on the Earth's surface from which the eclipse will *not* be visible.

4 Look at the diagram, which shows the Moon at four different positions in its orbit around the Earth.

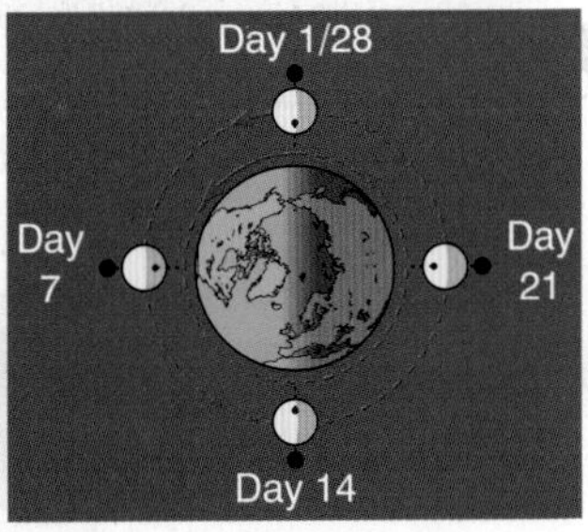

Day 1/28
Day 7
Day 21
Day 14

  a On which day will an observer on Earth see a full Moon?
  b On Day 14, what will an observer on Earth see when they look at the Moon?

5 Read the following statements. Copy out the correct statements, and correct the incorrect ones.
A: In the summer, the Sun rises higher in the sky.
B: In winter, it is colder because the Earth is farther from the Sun.
C: In December, the southern hemisphere is tilted away from the Sun.
D: Summer shadows are longer because the Sun is higher in the sky.

KEY WORDS

asteroid   atmosphere   axis   comet   eclipse
luminous   orbit   penumbra   phase   reflect
satellite   season   spin   umbra

Heating, cooling (8I)

1 Look at the key word list. Find two words which are opposites of each other. Explain their meanings.

2 Copy and complete the table using words from the key word list.

Key word	Definition
	a measure of how hot something is
	an instrument used to measure this quantity
	the scientific scale used for measuring this

3 Copy the table, and put suitable values of temperature in the first column. Don't forget the units!

Temperature	Description
	freezing point of pure water
	comfortable room temperature
	body temperature
	a hot drink
	boiling point of pure water

4 The answers to these questions can be found in the key words list.
  a What do we call energy moving from a hotter place to a colder place?
  b What do we call energy travelling from hot to cold through a solid?
  c What do we call a material through which energy can travel only very slowly?

5 We can use the particle model to explain why a solid expands when it is heated. Which is the correct reason?
  ● Its particles get bigger when they are heated.
  ● Its particles vibrate more and so take up more space.
  ● The air between the particles expands.

6 Now use the particle model to explain how heat energy conducts through a solid.

7 Which materials are the best conductors of heat energy?

> **◄ KEY WORDS ►**
>
> celsius scale   conduction   contract   convection
> expand   heat energy   insulator   particle model
> radiation   temperature   thermal conductor
> thermometer

Magnets and electromagnets (8J)

1 Copy the following sentences, choosing the correct words:

A magnetic material is one which is **attracted/repelled** by a magnet. Examples are **steel/nickel/tin/iron**.

2 Name a device which uses a magnet to help you navigate around the Earth.

3 **a** Copy the diagram of the magnet.
  b Add arrows to the field lines.
  c What do we call the space around a magnet?
  d What do we call the points labelled N and S? What is special about them?
  e If the magnet was hung horizontally, what would happen, and why?

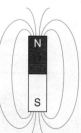

4 What is the difference between a permanent magnet and an electromagnet?

5 Draw an electromagnet and label all of its parts.

6 List all of the things which can make an electromagnet stronger.

7 Give two uses for an electromagnet. In each case, explain why an electromagnet is better than a permanent magnet.

> **◄ KEY WORDS ►**
>
> attract   compass   core   electromagnet
> magnetic field   magnetised   permanent magnet
> pole   repel   solenoid

Light (8K)

1 Which travels faster, light or sound? How could you prove this, during a thunderstorm?

2 Copy the table, and fill the first column with the correct words from the key word list.

Key word	Definition
	a very thin beam of light
	a source of a very narrow beam of light
	a picture of something, made from light
	all the colours of white light, separated out

3 Copy the table, and fill the first column with suitable adjectives. (Sorry, you won't find them in the key word list!)

Adjective	Definition
	won't allow light to pass through
	allows light to pass through, but a clear image is not formed
	allows light to pass through unaffected
	is a source of its own light
	left–right reversed

4 Copy the diagram and label or mark the following:
- the incident ray and the reflected ray
- the angles of incidence and reflection
- the normal
- two equal angles
- an angle which is 90°

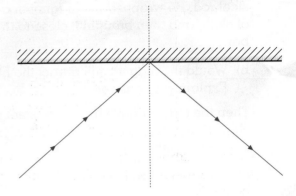

5 Draw a similar labelled ray diagram to show what happens when a ray of light strikes a glass surface. Which angle is greater, the angle of incidence or the angle of refraction?

Sound and hearing (8L)

1 Copy these sentences, choosing *all* of the phrases which are correct:

A louder sound has:
more energy
more vibrations per second
a higher pitch
a higher frequency
a greater amplitude

A higher sound has:
more energy
more vibrations per second
a higher pitch
a higher frequency
a greater amplitude

2 What is the unit of frequency? What is its symbol?

3 What is meant by 'range of hearing'? How does our range of hearing change as we get older?

4 Look at the three traces (A, B, C) which represent the vibrations of three sounds.
- **a** Compare A and B. What is the same for these two sounds, and what is different?
- **b** Compare B and C. What is the same for these two sounds, and what is different?

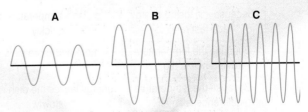

5 Give one word which means 'unwanted sound'. Why is this type of pollution hazardous?

1 Pete was investigating a bar magnet. His teacher gave him a second steel bar and asked him to find out if it was a magnet.

a) Describe how he could do this. (2)

Pete found that his magnet attracted the steel rod. He wrapped the magnet in a piece of cloth, and then brought it close to the steel bar.

b) Would the magnet still attract the bar? Explain your answer. (2)

Then the teacher gave him some steel paper clips. Some were plastic-coated. Pete put the magnet in with the paper clips.

c) What would you expect to happen? (1)

2 Imagine that you take a hot dish out of the oven. It cools down gradually before you serve the food from it.

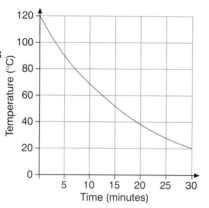

The graph shows how the dish's temperature changes as it cools down. Use your understanding of heat energy to explain why the dish's temperature shows this pattern.

The sentences below have been divided in half. Match up the correct halves, and then put them in the correct order to explain the shape of the graph. (6)

The temperature of the dish is much greater than . . .	its temperature is a few degrees above its surroundings.
The dish's temperature falls . . .	the temperature of its surroundings.
Because it is losing energy at a fast rate . . .	the dish's temperature falls quickly.
Because the temperature difference is big . . .	more and more slowly.
Because the temperature difference is small . . .	energy leaves the dish more slowly.
When the dish is cooler . . .	heat energy leaves the dish quickly.

3 Pip invented an electrical puzzle for her friends. It was a box with two switches and a light bulb. The box was connected to a battery.

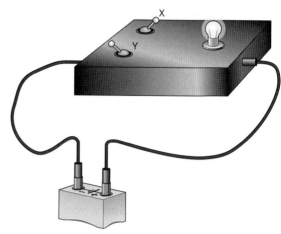

When the various switches were opened and closed, the light went on and off. The table shows the results.

X open	Y open	Light off
X closed	Y open	Light on
X open	Y closed	Light on
X closed	Y closed	Light on

a) Draw a diagram to show how you think the battery, bulb and switches were connected together inside the box. (3)

b) Pip wanted to add a second bulb which would come on whenever switch X was closed. On your diagram, mark with a letter **B** where you would add this bulb. (2)

4 Benson went on a fairground rollercoaster ride. The table shows the different stages of the ride.

		waiting at the top of the slope
		speeding up down the first slope
		moving back uphill
		speeding up again, on the second slope
		slowing down when the brakes came on
		stopped at the end of the ride

a) Copy the table and indicate the following points in the first column: (5)

   i) Mark with an X each of the two stages when the car had no kinetic energy.

   ii) Mark with a Y the stage when the car had most potential energy.

   iii) Mark with a Z each of the two stages when the car's kinetic energy was increasing.

b) What force caused the car to speed up down the slope? (1)

c) The car had less energy at the end than at the start. Into what two forms of energy had its energy been transformed? Choose from:

   sound   kinetic   heat
       potential   electrical   (2)

5 Look at the diagram of the ear.

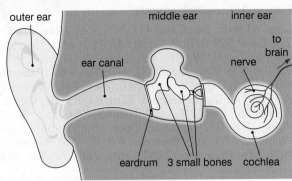

a) Which part is first to vibrate when a sound wave enters the ear? (1)

b) If the sound becomes softer, how will the vibrations change? (1)

c) If the sound becomes more high pitched, how will the vibrations change? (1)

Answers

1 a) Bring one pole of magnet up to either end of rod. If one end is repelled, it is a magnet. (2)

   b) Yes. Cloth is non-magnetic, so magnetic force passes through. (2)

   c) Clips attracted (including plastic-covered). (1)

2 The temperature of the dish is much greater than ... the temperature of its surroundings.

   Because the temperature difference is big ... heat energy leaves the dish quickly.

   Because it is losing energy at a fast rate ... the dish's temperature falls quickly.

   When the dish is cooler ... its temperature is a few degrees above its surroundings.

   Because the temperature difference is small ... energy leaves the dish more slowly.

   The dish's temperature falls ... more and more slowly. (6)

3 a) Two switches in parallel; battery and bulb in series with them. (3)

   b) Bulb B next to switch X. (2)

4 a)

X, Y	waiting at the top of the slope
Z	speeding up down the first slope
	moving back uphill
Z	speeding up again, on the second slope
	slowing down when the brakes came on
X	stopped at the end of the ride

(5)

   b) Gravity (1)

   c) Sound   Heat (2)

5 a) Eardrum (1)

   b) Smaller (i.e. smaller amplitude) (1)

   c) More vibrations per second (i.e. greater frequency). (1)

Acknowledgements

Picture Acknowledgements

Corel 61 (NT): 54t; Corel 417 (NT): 92b; Corel 456 (NT): 27t; Corel 539 (NT): 3; Corel 546 (NT): 46m; Corel 602 (NT): 6; Corel 709 (NT): 46t; Corel 771 (NT): 91, 90b, 92t, 93, 94t, 95t, 95m; Corel 772 (NT): 90t, 94bl, 94br; Corel (NT): 4, 25; Digital Vision 1 (NT): 74m, 74bl; Digital Vision 3 (NT): 101m; Digital Vision 6: 82m, 84, 85, 87; Digital Vision 7 (NT): 86t; Digital Vision 9 (NT): 83, 86m, 86b, 89; Digital Vision 12 (NT): 107; Digital Vision 15 (NT): 56tl; Digital Vision SC (NT): 99b; Photodisc 4 (NT): 75bl; Photodisc 6 (NT): 27b; Photodisc 17 (NT): 75br, 76bl, 77, 99t, 106;

Photodisc 18 (NT): 100, 101t; Photodisc 37 (NT): 48; Photodisc 54 (NT): 82b; Photodisc 67 (NT): 103; Photodisc 69 (NT): 56tr; Stockbyte 29 (NT): 74br; Alamy: Holt Studios International: 26; Martyn Chillmaid: 49, 78.

Every effort has been made to trace the copyright holders, but if any have been overlooked, the publishers will be pleased to make the necessary arrangements at the first opportunity.

Picture research by Stuart Sweatmore and John Bailey.

How to order Scientifica

Title	ISBN	Price £	IC/A please tick	✔	Firm Order Qty	£
Pupil Book 9 (Levels 4–7)	0 7487 7996 5	10.00	IC			
Pupil Book 9 Essentials (Levels 3–6)	0 7487 7997 3	10.00	IC			
Teacher Book 9 (Levels 4–7)	0 7487 8000 9	45.00	A			
Teacher Book 9 Essentials (Levels 3–6)	0 7487 8003 3	45.00	A			
Teacher Resource Pack 9	0 7487 8035 1	100.00	A			
ICT Power Pack 9 (eligible for eLC funding)†	0 7487 8037 8	300.00 +	A			
Assessment Resource Pack 9	0 7487 8040 8	65.00	A			
Online Test and Assessment 9†	0 7487 9488 3	100.00 +	A			
Special Resource Pack 9	0 7487 9205 8	150.00	A			
Workbook 9	0 7487 9186 8	3.50	IC			
Scientifica Presents People Like Us Reader 9	0 7487 9015 2	6.50	IC			
				Post and Packaging		£2.95*
				Total		

* Inspection or approval copies will be delivered carriage free to educational establishments. £2.95 post and packing will be charged on all firm orders.

† This electronic product is priced according to school size. To find out the price for your school please call Customer Services on 01242 267273

Prices valid until 31–12–05 – but are subject to change without notice

IC – Inspection Copy   **A** – On Approval   **+** – Plus VAT

Inspection/Approval copies are available to recognised educational establishments for evaluation free of charge for up to 30 days.

For terms and conditions please contact Customer Services on 01242 267287.

To request an inspection/approval copy, or to place a firm order, please complete and return this card, quoting an Official Order Number if required.

Official Order Number _____   To ensure your order is not delayed please check whether this is required by your establishment.

Name _____

Position _____

School/College _____

School/College address _____

_____

Postcode _____   Email _____

Telephone _____   Fax _____

Please send me a Science Catalogue ☐   Please send me an E-Learning Catalogue ☐   QCBZ

To contact your local educational representative call our dedicated support line on 01242 267284

Please photocopy and send back to:

Customer Services, Nelson Thornes,
FREEPOST SWC0504, Cheltenham, Glos. GL53 7ZZ

t: 01242 267273   f: 01242 253695   e: science@nelsonthornes.com   w: www.nelsonthornes.com

nelson thornes